#Fail:

Why the US Lost the War in Afghanistan

#Fail:
Why the US Lost the War in Afghanistan

Jonathan Owen

Blacksmith Publishing

Fayetteville, North Carolina

"In war many roads lead to success, and that they do not all involve the opponent's outright defeat." – Clausewitz

#Fail: Why the US Lost the War in Afghanistan

ISBN 978-0-9977434-7-0

Library of Congress Control Number: 2017962574

Printed in the United States of America

Published by Blacksmith LLC
Fayetteville, North Carolina

www.BlacksmithPublishing.com

Direct inquiries and/or orders to the above web address.

This book does not contain classified or sensitive information restricted from public release. The views presented in this publication are those of the author and do not necessarily represent the Department of Defense or its components.

While every precaution has been taken to ensure the reliability and accuracy of all data and contents, neither the author nor the publisher assumes any responsibility for the use or misuse of information contained in this book.

Contents

For the leaders of America's Armed Forces.
Godspeed in all your endeavors!

Foreword

In 2010, I concluded an exhaustive study on the US military's counterinsurgency (COIN) strategy. It identified the factors I warned would lead to major strategic failures in Afghanistan and Iraq for the United States and NATO. In doing so, my research pinpointed the critical failure points of the US COIN strategy and how they could best be remedied. To date, the findings have not been adequately appreciated or effectively addressed. As a result, the failures I predicted have come to pass. Still, there remains opportunity to rectify these failures.

When I initially published, the findings were so overwhelmingly clear, yet so divergent from establishment group think that my work was initially met with skepticism. A senior RAND expert who peer reviewed my research told me, "Joe, I can't believe we missed something that obvious." Indeed, it is troubling just how much false and inaccurate information has been allowed to pass as "fact" within military doctrine. If just a minimum of analysis and scrutiny was applied to the US COIN doctrine, the majority of it would be eliminated as counterproductive or ineffective. Consider the very basic "arm, train, equip, and advise" strategy. It's a foundation of US military COIN operations. Yet, those on the ground know, it has no correlation to counterinsurgent victory whatsoever. In fact, the only certainty of that strategy is a near 100% probability of counterinsurgent failure. Even worse, neither the military nor our intelligence community bothered to point out this very basic discrepancy to policy makers or the public. It's unclear whether it's incompetence or a lack of integrity. However, this information was easily verifiable. Regardless, this strategy is what the United States has blindly relied on in Iraq and Afghanistan. With this in mind, it was easy to predict the US military strategy in Afghanistan and Iraq would not work. Since then, Generals unwilling to admit they had been duped into a flawed strategy, have retired, resigned or been convicted of mishandling classified material. This creates a rare opportunity to repair the damage as many in Congress, Special Operations, academia, and the Intelligence Community are applauding the research as a revolutionary review of how the US should effectively conduct COIN operations.

Today the situation has changed. No one is arguing whether or not our military strategy will work. It is now broadly acknowledged that the US military's COIN strategy employed in Afghanistan was not only ineffective, but it is not and never was capable of achieving a decisive victory against an insurgent force. Symptomatic of this failure is the fact no one can articulate what is our specific strategy, mission, or end state in Afghanistan, even after over a decade and a half of war. Even among senior members of our military, officers will state that victory is not possible because no one knows what that looks like! This defeatist groupthink defies all logic and reason. It is not only inexcusable, but is pure intellectual dereliction by the most senior echelon of our military and political leadership. Thus, a new strategy, not a rehashing of previously failed COIN strategies, is absolutely necessary and long overdue.

The research contained in this book provides our leaders and war fighters with a new strategy that will rapidly achieve a decisive victory in Afghanistan. Yes, contrary to what you have been conditioned to believe by the naysayers that should have been fired years ago, clear-cut victory is possible. The strategy laid out in these pages is articulable, achievable, and easily understandable from the generals down to the privates. It provides a solid plan of action, not a theoretical outline, which is ready to implement. Further, the effectiveness of this COIN strategy is backed up by hard data and analysis of more than a century of named insurgencies. The evidence definitively shows the Taliban are far from being the toughest enemy the United States has ever fought and still can be easily defeated. Armed with this research, the new leadership in the White House and Pentagon has a unique opportunity to employ the study's findings and win. This requires a fundamental, yet highly feasible and easily implemented change to our COIN strategy. If implemented, it would save lives, cut costs, and rapidly bring the war to a successful close. Doing so now, even at this late stage in the conflict, can still yield strategic victories in Afghanistan, Iraq, and anywhere COIN is being waged.

The following book is a republished version of my original COIN study concluded in 2010. Only the book title, headings, and introductory portions have been updated. As such, the reader should find the text and conclusions, written in 2010, a prescient foretelling of what has now come to pass. The study is inherently written in an academic format to include the original executive summary. This was necessary so the research could be clearly laid out, verifiable, and cross-referenced with other sources. It is my goal to see the findings of this research reach the widest possible audience amongst our civilian and military leadership so a decisive victory can still be achieved in Afghanistan. Let not the lives of those lost be in vain.

Preface

"You can't kill your way to success in a counter insurgency effort.
— James G. Stavridis

General David Petraeus says violence in Afghanistan is likely to get worse before it gets better; however, it doesn't need to be this way. There are alternatives, but, so far, they have not been addressed. Why? It is because doing so, means admitting that the current strategy is fatally flawed. Unfortunately, admitting this growing reality has proven beyond the capability of our senior strategists. The results of this research explore one of these alternatives - border security.

There is overwhelming evidence to prove effective border security is the missing key, critical element in US policy, strategy, and doctrine for defeating cross-border, state supported insurgencies such as those faced in Afghanistan. Specifically, if the US does not make securing the Afghanistan-Pakistan border a top mission priority, the US military's counterinsurgency strategy in Afghanistan will face imminent failure. Just as important, and even more contradictory to prevailing group-think, is the finding that implementation of effective border security is highly feasible. The feasibility of effectively implementing border security is demonstrated by a strong historical precedent predating the classical Roman era and continuing through present day; especially, in rugged mountainous regions or open deserts such as found in Afghanistan.

Findings – The findings of this study should send a clear warning to senior policy makers as the results call into question the currently accepted and overly relied on tenets of COIN. They show effective border security plays a decisive role in defeating certain types of insurgencies, and is far more effective than other highly touted COIN strategies such as security force assistance (which includes training, advising, and equipping foreign militaries and police), amnesty programs, human terrain mapping, and economic "soft power" initiatives, combined; although, all of these elements should still be considered necessary parts of a holistic COIN strategy.

To illustrate this superior and inferior relationship, the historical record demonstrates in cases where states successfully overcame cross-border, state supported insurgencies, effective border security was the consistent factor that brought about insurgent defeat. However, without effective border security, there is almost no context for success even after years of sustained kinetic operations coupled with initiatives such economic development and substantial military training and advising efforts. Note that there is a fundamental difference between a policed peacetime border and a militarized wartime border. This research is decidedly focused on a militarized border.

Research clearly shows rapidly implementing highly effective border security is feasible even in extremely mountainous terrain and over long distances. A sampling of successful examples from the historical record include the rapid closing of the mountainous border between Greece and what was Yugoslavia, the closure of the Algerian borders with Tunisia and Morocco, the isolation of the Confederacy during the American Civil War known as the "Anaconda Plan," and the construction of "Hadrian's Wall" and the limes by the Roman Empire. The most extreme contemporary example was demonstrated by the rapid closure of China's border with India and Nepal by the People's

Liberation Army (PLA). The PLA's decisive action ended effective Tibetan resistance launched from the most rugged mountains on Earth, the Himalayas.

There are no "effective" half measures in COIN. Data strongly suggests partial employment of border security has no decisive effect. In cases such as in Afghanistan and Pakistan where the situation is still unstable and lifestyles are highly reliant on corruption and illicit/black market cross-border trade, local forces should not be relied on to independently and/or autonomously secure the border. Under these conditions, a plan reliant on locals to effectively operate as border security independently and/or autonomously generally fails and undermines overall strategy.

To overcome this, anti-corruption measures to include pairing of local forces with outside counterinsurgent forces is much more successful and a key area where SFA can be of assistance for establishment of long term security. Further, the conduct of limited, cross-border raids against insurgent sanctuaries are not only ineffective, but counterproductive. Historical precedent demonstrates that these actions are simply too limited to have any large scale decisive impact. Specific to the later point, any tactical gains achieved by limited, cross-border raids are typically far outweighed by the strategic costs. These raids generate a popular backlash throughout the raided country, which provides the insurgent a propaganda victory. This outrage amongst the local population and diaspora bolsters insurgent support and solidifies sanctuary. The effect of US drone strikes in Pakistan is arguably a testament to this effect. Even though the strikes have been highly effective in their lethality against senior Al Qaeda and Taliban leadership, they have been far too limited to decisively alter the course of the war and left the US viewed by the majority of Pakistanis as the gravest threat to their country. In short, if the counterinsurgent is not prepared to conduct a full-scale military invasion of a border nation, it should not attempt cross-border raids for anything less than an operation designed to bring an immediate and decisive end to the war.

Data consistently demonstrates a strong link between successful COIN strategies against cross-border, state supported insurgencies and the employment of effective border security. Further, the data passes a cross-check that links COIN failure with strategies that did not include effective border security. In fact, the consistency of the data shows the employment of effective internal and/or external border security against insurgents follows a typical progression with very predictable results as shown below:

1. A proto-insurgency begins that at minimum, has the potential for state sponsorship in a neighboring country.
2. The counterinsurgent begins small scale policing type operations against the "bandits."
3. The counterinsurgent is frustrated by lack of success and the growth of the insurgency.
4. The counterinsurgent begins large scale operations against the insurgency.
5. The insurgents retreat and establish/develop a cross-border, state supported sanctuary.
6. The counterinsurgent is repeatedly frustrated in its attempts to decisively engage insurgents, which repeatedly escape being cornered by withdrawing across borders.
7. The counterinsurgent attempts other more holistic COIN tactics such providing government concessions,

economic aid, and amnesty programs resulting in little success.
8. The insurgency continues to grow and evade decisive engagement.
9. The counterinsurgent in desperation finally acts to close the borders.
10. The insurgency launches large scale attacks against the border.
11. The insurgency suffers heavy and unsustainable losses attempting to breach the border.
12. The insurgency loses its ability to resupply and move its fighters to and from sanctuary.
13. The counterinsurgent begins to internally achieve decisive gains against the insurgency.
14. The insurgency typically within a month begins to show signs of critical instability.
15. The insurgency is decisively defeated, neutralized, or seeks a negotiated peace.
16. Political solutions are then achieved over a longer period of time with the state(s) that provided sanctuary to the insurgency, which effectively removes the sanctuary and inflicts a final decisive defeat on the insurgents.

Border security is a force multiplier and has effects much greater than just physical security. In fact, the effects of effective border security appear to compound themselves with time, which precipitates a tipping point against an insurgent force. In particular, border security is an extremely effective COIN tactic for the following reasons:

• Effective border security creates a self-reinforcing situation that overwhelms insurgents. Borders and barriers are a direct affront to insurgent legitimacy and critically hinder their ability to operate. As such, insurgents must attack and seek to breach effectively secured borders/barriers. This forces the insurgents to separate from the civilian population. This critical factor allows the counterinsurgent to decisively bring to bear his superior firepower and destroy the insurgents without causing collateral damage to the civilian population. The more desperate the situation becomes for the insurgent, the more critical it is for the insurgent to breach the barrier. The more they try, the more the insurgents are interdicted or killed. In all cases analyzed, this quickly became an unsustainable situation for the insurgent, which directly led to a negotiated settlement and/or collapse of resistance.

• Effective border security aligns current military strategy with a classic Clausewitzian focus on the fundamentals of defeating the enemy center(s) of gravity via attacking critical vulnerabilities. In fact, effective border security strikes directly at the center of gravity of a cross-border insurgency by turning the cross-border sanctuary into a critical vulnerability that can be cut and exploited. The insurgent can't influence populations if he doesn't have access to them, and he can't leverage access, or even gain sustained access, if he doesn't have the means to train, equip, and supply forces. For a homegrown, local insurgency, the people may be the source of basic military needs, but for a cross-border insurgency, freedom of movement and lines of communication are essential to exist. Therefore, isolation is necessary to decisively defeat an insurgent force.

• Implementing border security provides opportunities to enact key aspects of COIN such as consistency and commitment to the civilian population. Further, the construction of physical barriers provides needed jobs and processes through which the government can exert its influence and build rapport.

• Effectively controlled borders foster licit trade networks while limiting illicit networks. The existence of illicit trafficking networks directly challenges the legitimacy of the government and must be removed over the long term to bring about a stable functioning state. This stifling of the black market, specifically in the traffic of weapons and drugs, directly undercuts the funding for many insurgent groups, warlords, and criminal elements while boosting the legitimacy of the government. It also cuts needed supplies of weapons, ammunition, and equipment for insurgent forces and makes maintaining insurgent lines of communication much more difficult.

• Implementing effective border security to include a system of static and mobile defenses in depth allows stronger state entities such as the United States to effectively leverage their vastly superior materiel resources, heavy construction expertise, command and control technologies, and mobility capabilities toward decisive gains. This is noteworthy in that COIN offers few opportunities to leverage inherit strengths of a state against an insurgent.

Removing misconceptions about border security is critical for policy makers and military leaders. The facts clearly show securing long mountainous borders and desert regions such as those found in Afghanistan can rapidly be accomplished and have a strong historical precedence. Mountains naturally channel movement and can be controlled by holding finite strategic ground supported by a mobile reserve and indirect fires. Further, sparse vegetation at various altitudes and low population density also adds to the ease of identifying illicit cross-border movement. Desert regions are easiest to secure and are well suited to a fluid mobile defense-in-depth, which the United States is well suited to conduct.

Combining this with modern, highly sophisticated intelligence, surveillance, and reconnaissance assets, the US is well poised to secure the Afghanistan-Pakistan border should it choose. Using the highest average per mile cost of the most sophisticated (and proven) integrated border security systems fielded in the United States, Europe, Saudi Arabia, and Israel, the Afghanistan-Pakistan border could be secured for as little as $3.3 billion dollars. To put this in perspective, it is about the same amount of money the US was spending approximately every two weeks on the war in Iraq at its height, it is roughly the same amount the US has invested in the Marine Corps Expeditionary Fighting Vehicle to date, and it is very close to the amount allotted for the infamous "Cash for Clunkers" car buy-back program. Considering the potential costs versus benefits of actually winning the war, this is a very small price to pay and makes both sound fiscal and strategic sense.

To be sure, border security alone is not alone sufficient to defeat an insurgency. Although the data is overwhelming in that a war against a cross-border, state supported insurgent is almost unwinnable without effective border security, winning a COIN struggle requires more than just military might. For example, effective SFA and other types of security cooperation will allow for an eventual turn-over of tasks and long term sustained stability after initial security is achieved. Considering this, building rule of law and effective government institutions, efforts to rebuild economies, anti-corruption efforts, and other aspects of what has been termed "soft power" will also continue to be a necessary part of a holistic COIN strategy.

Without immediately implementing a full-scale, unilateral initiative to seal the Afghanistan-Pakistan border, the US/NATO force will not be able to achieve decisive results against the Taliban and Al Qaeda insurgents. This will lead to the imminent exhaustion and defeat of US/NATO forces. However, the following findings and recommendations provide a roadmap, which if implemented, will provide the US with the first real path to a decisive victory over the Taliban and can also be used to mitigate asymmetric, "fourth generation" threats posed by insurgent forces globally.

Recommendations – It is not too late to win. The current US/NATO COIN strategy in Afghanistan, which does not isolate the insurgency from its sanctuary, will be insufficient to achieve the domestic security necessary to neutralize the threat posed to Afghanistan and the region by Taliban and Al Qaeda insurgents operating from Pakistan. Even as this study is being released, Taliban elements are spreading out from Pakistan across Afghanistan and are linking with Uzbek and other Islamist insurgent groups. These elements are now coordinating attacks from within Afghanistan against Uzbekistan. This should be clear evidence this insurgency will spread if not stopped at the Afghanistan-Pakistan border and will continue to have long range security implications for the US. Successful COIN in the Afghan War will therefore require a radical change in strategy that includes a concerted effort and the requisite retasking of assets to seal the Afghanistan-Pakistan border.

The counterinsurgent force must remove access to and from the cross-border sanctuary in Pakistan. Although gradually increasing the percentage and responsibilities of Afghan forces in support of this will be necessary and beneficial, the US/NATO forces must plan on initially unilaterally undertaking and accomplishing this task. Failure in this regard will allow for the influences of corruption on local indigenous forces to open holes in the border that will be exploited by insurgents and undermine the overall effectiveness of the plan.

As important, the US cannot rely on Pakistan as the answer. Pakistan is the problem. Waiting for Pakistan to address the Taliban problem is akin to waiting for the Soviets to address the communist problem. Policy-makers that believe Pakistan will unilaterally act against its own self-interests for the benefit of the US are dangerously mistaken. The aid we provide Pakistan only ensures the Pakistani government will work through covert channels to keep the insurgency alive so that the foreign aid continues to flow. Further, the aid will not be used to decisively defeat insurgents that provide strategic depth and a ready source of fighters for contingencies and guerilla warfare against India. In Pakistan's defense, it may well be true that destroying the Taliban and Al Qaeda elements within the country is simply beyond its capabilities and there is good argument to support attempts at this could destabilize the country to the point of collapse.

Precipitating the collapse of a nuclear power is certainly not in the best interests of the US. Consider this in its totality, relying on Pakistan to fight our war will not be a viable strategy. If however, the aid was recapitalized, this aid would be more than enough to cover the costs of constructing a comprehensive and effective border security system that would decisively impact the outcome of the insurgency in favor of the US/NATO. Once effective border security is established, the conditions and calculus that made the Taliban an important national security element for Pakistan to maintain, will fundamentally change. This pattern has been repeated in nearly all case studies and if the precedent holds true for Pakistan, once the insurgents are unable to operate in Afghanistan, the Taliban will become more of a nuisance than an asset to Pakistan. This will prompt Pakistan to actually take decisive action against the Taliban that at minimum would include ceasing all support. At this juncture, the US can politically reengage Pakistan to help decisively end the conflict from a position of mutually aligned national interests. Only after the benefits Pakistan provides the Taliban and the Taliban provides Pakistan have been mitigated through border security, can a long term, and final political decision that decisively ends the insurgency, be achieved.

SFA in the form of weapons (small arms to heavy weapons and aircraft), training, and provision of advisors does not have decisive effects. In fact, although the US/NATO strategy in Afghanistan has placed high priority on SFA and pegged its hopes of victory on standing up a trained local military and police force, there is little historical evidence to suggest this is a game changer. For example, the Soviets managed to train, equip, and deploy a large conventional Afghan military that included air, armor, intelligence, Special Forces, and police units along with conventional ground forces to no decisive effect. Even worse, there is substantial and growing evidence to suggest that massive SFA provided by the US has fueled corruption, which undermines the government and has provided the insurgency with a steady stream of well-armed and trained recruits and spies. However, in defense of SFA, there is equally strong evidence to support that it is a force multiplier when coupled with effective border security. In particular, it supports sustaining long term stability and effective hand-over of mission tasks to local forces. As such, conducting SFA should not be ruled out, but must be understood as a supporting, not a primary, focus of any COIN strategy.

The closure of the border would allow the US/NATO to consolidate its gains and maintain pacified areas. This would actually accelerate US/NATO gains and allow for the first time, decisive actions to occur. Implementation of a secure Afghanistan-Pakistan border would be devastating to the insurgency because the majority of tier one Taliban fighters (principle agents conducting the war) are foreigners and come from cross-border areas to incite the local populations (tier two Taliban) to arms. Typically, tier two fighters won't act without the foreign catalyst and the population certainly would feel more secure working with US/NATO forces. It would also reign in warlords that challenge the government's legitimacy financed by illicit cross-border smuggling operations. This would include curtailing the drug trade and stifling the movement of bomb making materials, weapons, and ammunition. Further, leveraging locals to help construct a physical barrier would provide critically needed jobs and a process through which the government and legitimate leaders could exercise authority. The US, more than any other country in the world, excels at large scale construction and engineering tasks. Building a physical border allows the US to decisively put its conventional strengths to work in a

COIN environment. As well, for the first time, the US has in-hand an effective strategy that not only can win the war, but allow it to regain the initiative against insurgent-style warfare worldwide.

The research shows effective border security has policy implications beyond Afghanistan and could be used to effectively address a broad range of issues, such as WMD proliferation, illicit smuggling (weapons, people, drugs), and trans-national terrorism. Expanded research shows effective border security is a highly effective counterterrorism tool and has proven to be more individually valuable in stopping the incidence of terrorist attacks, than the use of intelligence, military operations, or targeted killings. Border security initiatives also would align better with conditions that have historically been more suitable for United Nations involvement, which could relieve the US of at least some demands in manpower and capital expenditure. Border security is also a more politically palatable and effective tool than much riskier strategies, such as covert action and overt military operations, including extensive occupation. Further, border security operations are low risk for US personnel, are unlikely to cause any collateral damage, could be used to bolster a nation's economy, and would on the whole be much more affordable than many other options. For the United States, this could become the primary policy strategy for dealing with "politically denied areas."

The research demonstrates the need to fundamentally revise US military strategy, doctrine, planning, and training in regards to COIN. Border security is a key critical element to COIN success so military doctrine, specifically Field Manual (FM) 3-24 Counterinsurgency, will need to be revised. For example, instead of an overreliance on SFA doctrine focusing on supporting police and military units, it should be revised to target the development of strong border security forces that operate with inherent checks and balances to prevent corruption. General purpose forces (GPF) should also be trained to implement and operate physical border security. This would include construction of a defense in-depth that contains fixed positions on key terrain supported by pre-sited indirect fires and foot, vehicle, and helicopter mobile quick reaction forces that leverage the latest in sensors, and intelligence, surveillance, and reconnaissance (ISR) assets. Further, at the senior level, planning for a campaign must include an assessment of the potential for insurgency. In the event conditions exist that would suggest a high probability of an insurgency leveraging cross-border sanctuary developing, the military must plan for extended, comprehensive border security. This will be a paradigm shift in military strategy and planning and precipitate the need for training of specific tactics, techniques, and procedures for border security.

The results of this research come at a decisive juncture in the US/NATO war in Afghanistan. Should the current course be continued, the US/NATO forces will face a humiliating military defeat and foreign policy disaster with the potential for dire national security implications at home. Should the US/NATO force incorporate the findings of this research in a revised strategy, it will have a decisive impact in defeating the insurgency and establish the conditions for Afghanistan to become a sovereign, stable, and functioning state within the current 2014 timeline. The findings fundamentally change the understanding of the way COIN strategy is developed and have revolutionary implications for US security policy. The results prove that a key critical element has been missing from US COIN strategy and doctrine for more than a hundred years (arguably the last deliberate COIN use of border security by the US military was the "Anaconda Plan" implemented during the America Civil War). Armed with the knowledge that effective border security is a critical and necessary element to defeat cross-border, state supported insurgencies, the US can turn its conventional strengths in firepower, industrial might, technology, and manpower toward physically securing the Afghanistan-Pakistan border. Only through this action, will the US/NATO force turn the tide against asymmetric threats, and once again become victorious in insurgency warfare.

This book is dedicated to the tens of thousands of men and women killed and wounded in Iraq and Afghanistan that selflessly answered their nation's call to duty. May their collective sacrifice not be in vain.

I would also like to thank those of you that supported me throughout my research; Kimberly, Craig, Dan, and Josh.

Introduction

"Let me work here for Britain's sake – at any task you will – A marsh to drain, a road to make or native troops to drill. Some Western camp (I know the Pict) or granite Border keep, Mid seas of heather derelict, where our old messmates sleep." — *From Vectis to Hadrian's Wall*, Rudyard Kipling

Study Objective

This thesis project will test the importance of effective border security in counterinsurgency (COIN) campaigns, specifically those aimed at defeating state supported transnational insurgencies. It will seek to determine whether border security is just one of many elements of a successful COIN strategy—as it is typically characterized (if at all) in both COIN manuals and studies of COIN–or whether it is in fact a key *critical element necessary for neutralizing and eventually defeating these types of insurgencies* that operate from cross-border sanctuaries and receive at least some level of host state support.

Policy Relevance

The conclusions of this thesis are intended to have general impact on the study of COIN and specific policy implications for the US/NATO strategy in Afghanistan where US/NATO forces are directly engaging in a COIN campaign against insurgent forces benefiting from cross border sanctuary and at least minimal state support from elements within the Pakistan government. If a positive correlation is found between effective border security and the success of a COIN campaign, the viability of the current COIN strategy in Afghanistan will be challenged. The failure thus far to secure the Afghan-Pakistan border will represent a critical oversight in the US/NATO strategic plan--one that strongly suggests the COIN mission will suffer near certain failure without a major policy adjustment. Further, a fundamental overhaul of military COIN strategy, doctrine, planning, and training across the doctrine, organizational, training, materiel, leadership, personnel, facilities, and policy (DOTMLPF-P) spectrum will be necessary, as the concept of border security is almost completely absent in contemporary US military strategy (this is discussed in greater detail in this study's literature review). On the other hand, if no such correlations are found between effective border security and successful COIN, it would refute arguments suggesting the Afghan War is not winnable because the insurgency enjoys vast safe havens in Pakistan. This should bolster the case for the efficacy of the current US/NATO strategy. Finally, the results of this study will have implications for how the US views the costs and benefits of foreign intervention, and how it designs its strategy for dealing effectively with transnational terrorism and insurgency.

Background

Nearly daily reports can be found discussing issues related to the Afghanistan-Pakistan border. These reports make it clear that the Taliban and Al Qaeda fighters are reaping large benefits from this cross-border sanctuary and as a result are able to continually undermine the ability of NATO to achieve a secure and stable Afghanistan. For example, a recent article titled, "NATO Can't Stop the Flow of Bomb Materials from Pakistan," dated October 2010 discusses NATO's inability to stop the flow of bomb making material into Afghanistan from Pakistan.[1] It highlights how Pakistan is unwilling and in some degree also unable to do anything about this situation.[2] Thus, one can deduce insurgents operating from sanctuary in Pakistan are undermining NATO efforts in Afghanistan and that Pakistan is unwilling and or unable to do anything about it; one should also question what is being done by NATO to stop the problem. Implementing effective border control seems to be the logical answer, but no large scale systematic effort appears to have been undertaken. In fact, it seems as if NATO has generally abdicated the responsibility for securing the border to the Afghan Border Police (ABP), which it openly admits is unable to effectively conduct its mission.[3] This limited assistance is even questioned by policy experts, as it hinders NATO operations and US support.[4] The ABP, despite increased training and equipment provided by the US government, is simply not able to provide "effective" border security due to issues of corruption, limited resources, and failure to actively patrol the border.[5] This is an

---

[1] Dreazon, Yochi J. "NATO Can't Stop Flow Of Bomb Materials from Pakistan - Friday, October 29, 2010." *NationalJournal.com*. 27 Oct. 2010. Web. 15 Dec. 2010. <http://nationaljournal.com/nationalsecurity/nato-can-t-stop-flow-of-bomb-materials-from-pakistan-20101027>.

[2] Entous, Adam, and Siobhan Gorman. "White House Report Faults Pakistan's Anti-militant Campaign - WSJ.com." *Business News & Financial News - The Wall Street Journal - WSJ.com*. 6 Oct. 2010. Web. 15 Dec. 2010. <http://online.wsj.com/article/SB10001424052748703298504575534491793923282.html?mod=WSJ_hpp_MIDDLENexttoWhatsNewsThird>.

[3] Nelson, Soraya. "Afghan Border Police Make Progress, Slowly : NPR." *NPR : National Public Radio : News & Analysis, World, US, Music & Arts : NPR*. 5 Mar. 2009. Web. 15 Dec. 2010. <http://www.npr.org/templates/story/story.php?storyId=101377415>.

[4] Nelson.

[5] Nelson.

operational testament to the fact that although border security is repeatedly addressed as an aspect of COIN and even by senior policy makers such as US President Barack Obama, it is at best of a secondary, peripheral importance. Further, nearly all mainstream publications on the subject stop short of suggesting that to achieve success, this is a necessary condition, and rule out, if even mentioned, physically securing the border. This is predicated on the belief from senior policy makers that countries, often with borders that extend over a thousand miles, are simply unable to effectively secure them. Department of Homeland Security Secretary Janet Napolitano's recent statement, "You're never going to totally seal that border;" in reference to the US-Mexico border is indicative of this pervasive assumption, even though the Secretary likely made this statement with US domestic politics in mind, as much as physical border security.[6] However, this apparent mainstream assumption doesn't seem to hold up against more empirical research into cases where border security has been effectively implemented, such as in Israel, where it seems the counterinsurgent force has reaped large and immediate benefits.

Seeing the ill-informed and barely existent interest toward securing Afghanistan's borders, one should ask how serious a threat this cross-border issue poses to the COIN efforts in Afghanistan. At least anecdotally, it appears the threat posed by cross-border insurgents is indeed dire, noting the failure of the US in Vietnam and the Soviets in Afghanistan, against what most would

---

[6] Cassandra, Adam. "Napolitano: 'You're Never Going to Totally Seal That Border' | CNSnews.com." *CNS News* | *CNSnews.com*. 25 June 2010. Web. 15 Dec. 2010. <http://www.cnsnews.com/news/article/68494>.

have considered far inferior insurgent forces operating from cross-border sanctuaries. As such, this thesis sets out to analyze how effective border security may or may not affect a COIN effort against a cross-border state supported insurgent. The assumption being that border security is highly underappreciated in policy and strategy and may be an overlooked critical element to COIN.

Hypothesis

Hypothesis to be Tested: In state supported cross-border insurgencies, achieving effective border security is a critical and necessary—though not in and of itself a sufficient—condition for counterinsurgent victory.

In order to render insurgent operations militarily ineffective and ultimately neutralize the military threat they pose to the state, the counterinsurgent must employ an effective border security strategy, which entails interdicting or preventing insurgent cross-border movement (including the movement of personnel, equipment, and funding). Effective border security may not eliminate a cross-border insurgent sanctuary, but if the hypothesis is correct, it does predict it will effectively deny the insurgency the ability to benefit from that sanctuary inside the contested state. This may be accomplished via a variety of methods ranging from political agreements to physically closing a border. Either way, the ultimate end state must be that the insurgency is denied free movement to and from its sanctuary. This would imply the counterinsurgent must be prepared from the onset of hostilities to independently physically seal a border in the event other avenues such as political agreements fail. In the absence of effective border security, a cross-border insurgency will continue to pose a military threat to the state and indefinitely deny the counterinsurgent the ability to establish security and stability necessary for effective governance. This indefinite extension of hostilities is extremely dangerous to a non-indigenous counterinsurgent force primarily composed of foreign personnel due to the eventual exhaustion of projecting forces over time in political will, public support, and capital expenditures in money, materiel, and manpower, as witnessed by the Americans in Vietnam and the Soviets in Afghanistan. This is precisely the situation the US/NATO counterinsurgent force is facing in Afghanistan against an insurgency that freely moves to and from a cross-border sanctuary in Pakistan where it enjoys thinly veiled covert state support.

Terms of Reference

In this project, the counterinsurgent force will be deemed successful if it is assessed to have "neutralized" the military threat posed by an insurgency, rather than altogether "defeated" or "destroyed" the insurgent force. The threat posed to a state by a rebellion will be considered neutralized if the incumbent government is able to extend security and stability throughout its territory such that effective governance is possible. This will mean any remaining insurgents will be militarily ineffective (the term "military ineffective" is used to describe a force that may still retain the will to resist, but no longer has the means to conduct large scale military operations capable of destabilizing a government). Thus, this enables a desirable COIN end state where the insurgency is unable to conduct operations that seriously challenge the counterinsurgent's control over the state and its territory, even though limited terror attacks may persist for some time, as remaining pockets of insurgents are captured, killed, or converted. This lesser terminology is necessary because, although an insurgency may be rendered militarily ineffective in a contested country, it does not necessarily follow that the insurgency will be politically ineffective (although it is likely severely weakened), that no lethal attacks can be conducted, that every insurgent will be killed, captured, or converted, or that the insurgency will be defeated beyond the borders of the contested country.[7]

"Effective border security is necessary though not in and of itself sufficient" means even if a border could theoretically be completely sealed and the insurgent cut off from its sanctuary, the COIN strategy will still not be enough to achieve a victory unless it incorporates a host of other aspects to win over the population that include providing effective governance. However, this should not detract from the fact that the hypothesis still argues that dealing with a state supported cross-border insurgency, effective border security is a key, critical element that must be implemented achieve a COIN victory.

Border security is further considered "effective" if it denies the insurgency of manpower, materiel, funds, and the benefits of sanctuary found outside the borders of the counterinsurgent's country. The threshold for this

[7]In the case of Israel, for example, both Palestinian and Lebanese militant groups have been effectively contained by the borders of their respective host nation states. This does not mean, however, that groups such as Hamas and Hezbollah have been altogether defeated, or have given up resistance. It means that security has been achieved in Israel to such a degree that although Israel must remain vigilant in its defenses, neither group threatens the existence of the Israeli state.

is admittedly ambiguous, but can be identified by a marked reduction in insurgent activity over time, beginning with implementation of border security. If "effective" border controls are in place a continuously more desperate supply and manpower situation for the insurgent will be identifiable, and will continue until a general state of security is achieved, due to the inability of the insurgent to carry-on further military operations. If this hypothesis is valid, it is likely that this pattern will not follow a linear progression, but rather, will show acceleration over time in the degradation of the insurgent's fighting ability that reaches a threshold and then quickly culminates in a decisive end of insurgent activities.

The benefits of a sanctuary to an insurgency are immense, and when coupled with state sponsorship, include things such as weapons, equipment, intelligence, medical treatment facilities, training facilities, food, shelter, communications, and a base from which to plan, stage, and launch operations. According to the hypothesis, these benefits are so great, that without, an insurgency that resorted to moving to a cross-border sanctuary for operation will succumb to the counterinsurgent inside the contested state.

Following this logic, if these benefits are removed by cutting off the insurgent from his base of operation and supply, the insurgent will be forced to operate in areas already deemed incompatible with long-term resistance. The result will be the counterinsurgent will gradually gain the upper hand and ultimately be able to corner and eliminate any remaining pockets of resistance, which enables the incumbent government to establish security and return stability and governance to a region. Thus, cutting the insurgent off from this sanctuary is predicted to be a necessary precondition to neutralize an insurgency operating from cross-border sanctuary/safe-havens. If this proves correct, the insurgent isolation, both internally and externally, will set the positive security conditions for a state to stabilize its territory and reassert effective governance.

Research Methodology

The research for this project used a phased approach that began with a broad literature review of insurgencies and subsequently identified relevant case studies. Relevant case studies required a situation where an insurgency operates from a state supported cross-border sanctuary. Further, insurgencies evaluated survived the proto-insurgency phase long enough to take root, sustain, and develop into a generally recognized and named insurgency. Conflicts of this type, where no border security was implemented and the counterinsurgent force lost, provide the predicted baseline. If the tested hypothesis is to prove correct, this should be the norm, barring implementation of border security.

Conflicts where border security was implemented will provide confirmation if it can be shown they indeed contributed heavily to the counter insurgent victory. Conflicts that appear to be outliers (a situation where effective border security was implemented, but the counterinsurgent still lost, or cases where the counterinsurgent defeated a state supported cross-border insurgency without the aid of border security) will also be identified and addressed for conditions that disprove or modify the project's working hypotheses.

Testing the Hypothesis

This project will analyze data related to the impact of border security on insurgencies via the case study method in order to provide a measure of the importance of border security to successful COIN. If the hypothesis is correct, the data should demonstrate a direct correlation between effective border security being implemented, a reduction in insurgent cross-border activity (movement of personnel, materiel, and funds), and a subsequent reduction in the insurgency's ability to conduct effective military operations.

Best test case scenarios will strongly correlate border security with a decisive counterinsurgent victory. Strong indicators of the hypothesis' strength will be a direct correlation in time between implementing effective border security and measurable reductions in insurgent activity and capabilities. Lead indicators will include such metrics as persons turned back at a border or the amount of intercepted contraband; lag indicators will identify improvements in security due to reductions in insurgent military activity.

Literature Review

Throughout the large body of literature and research on insurgencies, it may initially appear that no two insurgencies are alike; thus, no strategy generalizations should or could be made. However, although the overarching term "insurgency" does include a huge variation in conflict forms ranging from localized terrorism to wide spread uprisings for independence, a closer examination reveals insurgencies can be categorized by their nature and generalizations can be made about the most effective strategies for both the insurgent and counter insurgent to pursue. This is evident in existence of some of the most famous insurgent literature from the likes of Mao Tse-tung and Che Guevara, counter insurgent works from Roger Trinquier and David Galula, as well as doctrine like the US Army's Field Manual 3-24 *Counterinsurgency.* However, in these works, state supported cross-border sanctuary is given little attention, specifically by the counter insurgent literature, forcing one to question whether or not this issue has relevance to COIN.

This lack of COIN strategy, related to border security and cross-border sanctuary may best be answered by a statement from military historian Martin van Creveld:

> The first, and absolutely indispensable, thing to do is throw overboard 99 percent of the literature on counterinsurgency, counterguerrilla, Counterterrorism, and the like. Since most of it was written by the losing side, it is of little value...[8]

Considering that van Creveld does have a point that the primary sources for counter insurgent literature were officers that fought on the losing side, such as Trinquier and Galula, it seems likely the body of literature related to effectively conducting COIN is still incomplete and should be viewed neither as static nor sacrosanct. As such, a literature review of military histories and strategy related to insurgencies that have a decidedly cross-border state supported nature was conducted in search of evidence supporting border

---

[8] Van Creveld, Martin, *The Changing Face of War: Combat from the Marne to Iraq*. New York: Ballantine, 2008. 268. Print.

security as a key missing link in current strategy and doctrine.

The body of literature on insurgency and COIN is substantial and spans both contemporary and historical works. The sources include news articles, journals, official doctrine, academic studies, and books. Although primary sources do exist, secondary sources tend to be the most prolific and readily available. In most sources on insurgency, some mention is made of border areas and or security. However, even though some studies call for a need to have better border security, and a few highlight its importance, there is a striking absence of sources that specifically address border security as a central and critical component of an effective strategy for neutralizing cross-border insurgencies. At best, sources will identify the removal of sanctuary as important, but then fall short of suggesting the primacy of a direct border security plan that would be a necessary course of action to deny insurgent sanctuary.

In fact, no source evaluated during the literate review identified an effective border security strategy as necessary for COIN victory, although plenty of sources are quite clear about the fact sanctuary was a key component of a successful insurgency campaign. This omission is curious and may be rooted in preconceived notions regarding the feasibility of border security, as well as a natural American aversion toward closed borders. There are numerous examples of insurgent victories that were dependent upon the availability of sanctuary including Laos and Cambodia during the Vietnam War and Pakistan during the Soviet-Afghan War. In fact, it appears most cross-border state supported insurgencies that survive past the proto-insurgency phase do succeed, except in cases where borders have been effectively closed. These cases seem to portray initially successful cross-border insurgencies that either collapsed after state supported sanctuary was removed, such as in the Greek Civil War and Tibet's War for Independence, or insurgencies that were severely curtailed or neutralized such as in Israel, Turkey, El Salvador, and Kashmir.

Academic Literature

RAND studies as a whole have formed the bulk of open source analytical literature on insurgencies. In particular, RAND has developed and coded extensive data sets on almost 100 insurgencies dating roughly since the end of World War II. The RAND studies contributing heavily toward the literature on insurgency with border security implications include:

Money in the Bank Lessons Learned from Past Counterinsurgency (COIN) Operations;
Pacification in Algeria; David Galula
Counterinsurgency in Afghanistan
How Insurgencies End
Insurgency and Counterinsurgency in Algeria
The Role of the Sanctuary in Insurgency: Communist China's Support to the Vietminh 1946-54

These studies provide extremely valuable data, identify sanctuary and state support as key elements in insurgent success, and provide detailed case study analysis. The 1964 RAND study on Algeria (*Insurgency and Counterinsurgency in Algeria),* directly addresses effective border security operations that successfully sealed the border between Algeria and Tunisia, effectively isolating the insurgent from its primary base of support. J.J. Zasloff's study, *The Role of the Sanctuary in Insurgency: Communist China's Support to the Vietminh 1946-54*, is a detailed case study focused specifically on the impact of state support and sanctuary for an insurgency.

The importance of the document can be summed up by a single quote buried within the text by Walter Lippman that states, "It is for all practical purposes impossible to win a guerilla war if there is a privileged sanctuary behind the guerilla fighters."[9] However, although this quote proves that even in the years preceding expanded US involvement in Vietnam, thinkers clearly had linked sanctuary with insurgent success, Zasloff applies an incredibly high analytical bar to the analysis and fails to link sanctuary with the success of the Vietminh. In fact, Zasloff almost totally dismisses the impact of the Chinese supplying the Vietminh up to 4,000 tons of supplies a month by 1954, on the French ability to ultimately prevail in the conflict. Zasloff argues that state supported sanctuary "does not demonstrate a casual relationship" to insurgent victory and dismisses sanctuary, as an inferior condition to other factors such as the insurgent will to fight. This failure to correlate the impact more strongly, by a heavily respected research institution such as RAND, almost certainly influenced contemporary thinking at a critical juncture in US policy, as it began to ramp up its involvement in the disastrous war in Vietnam.

Other more recent work such as Seth Jones', *Counterinsurgency in Afghanistan,* positively identifies the sanctuary and support provided by Pakistan as a critical element for insurgent success (or inversely the

---

[9] *Washington Post* 15 Sept. 1963. Print.

counterinsurgent's failure), but also is a victim of the "curious failure to correlate insurgent victory to a comprehensive border security plan." Specifically, Jones falls short of addressing the need for broad implementation of a border security strategy, and instead, recommends the watered down, policy neutral strategy, of attempting to politically convince the Pakistanis to better police their border. This ineffective approach (relying on politics to will the Pakistanis to act) has more than 30 years of history that conclusively demonstrates this strategy is beyond Pakistan's capability and is contrary to their vital state interests. *How Insurgencies End*, by Ben Connable and Martin C. Libicki perhaps do the most to definitively link state supported sanctuary with insurgent victory.

Their large-N study of insurgencies leveraging RAND's coded data on insurgencies shows a high correlation between insurgent victory and state supported sanctuary. Nonetheless, Connable and Libicki did not approach this study, and as such analyze the large data set, from a perspective of how effective border security affected the overall outcome of cross-border insurgencies. Initial review of the RAND data set seems to substantiate the general premise of the hypothesis that without effective border security, it is highly unlikely for a counterinsurgent to prevail over an insurgent force with a state supported sanctuary barring some major game changing unforeseen event such as a key death of a leader, internal coup, or fundamental change in the political situation. The data set also appears to correlate border security with neutralized insurgencies that are still on-going such as in Israel and India (Kashmir) where the counterinsurgent benefits from relative peace and reasonable security. This type of insurgency category (on-going) appears to have by far the highest proportional incidence of border security activity suggesting a positive correlation in favor of the counterinsurgent if the bar for winning is set at effective neutralization as the hypothesis contends. However, the data set is still incomplete as it misses insurgencies such as Turkey's on-going conflict with the Kurdish separatists and its coding still suffers from the affliction of all large-N studies in that it doesn't provide the detail of highly scoped case studies. Specifically, due to coding generalities, RAND's data correlates the Algerian War of Independence as an example of the failure of border security, when in reality; the French achieved a military victory over the Algerian insurgents due in large part to a highly effective border security strategy. Thus, it is critical that during the conduct of research, the coding of the large-N data set be validated through at least minimal case study.

RAND is far from the only academic source on COIN. Countless books address issues related to COIN and also provide insights into border security. For example, Colonel Thomas Hammes' book, "The Sling and the Stone," is hailed by the *Marine Corps Gazette* as "our generation's single-source, strategic document about fighting insurgencies, taking its place alongside the timeless tactical *Small Wars Manual*."[10] However, in spite of the valuable contributions this book makes toward the study of COIN, the author fails to make any connection between cross-border sanctuary and the need to effectively secure a border. In fact, even though the book spends considerable time discussing significant cross-border insurgencies such as Vietnam and both

[10] Thomas Hammes, *The Sling and the Stone: on War in the 21st Century* (Grand Rapids, MI: Zenith, 2006).

Soviet and current US/NATO wars in Afghanistan, the idea that borders and sanctuary may be a factor in COIN is absent. "The Accidental Guerilla," by David Kilcullen, presents another in-depth and valuable contribution to the body of COIN literature. Kilcullen provides remarkable insights related to the value and impact of cross-border sanctuary and state support to an insurgent in his case study on the Kunar Province of Afghanistan.[11]

Kilcullen specifically lays out four key factors provided by what he calls "active sanctuary" in Pakistan that include training and logistics support systems, political and religious leadership, the recruiting base for full-time fighters, and external sponsors and financial backers.[12] He is careful to note the full-time fighters are recruited from various locations, but generally trained across the border in Pakistan and then move within the Afghan population where they conduct political and religious indoctrination, armed propaganda, intimidation and killing of government supporters, collection taxes, and direct attacks on high value coalition and government targets.[13] These fighters are identified by Kilcullen as the primary actors in the insurgency and tend to be foreigners in the area they operate reinforced with legitimate foreign fighters such as Chechens, Tajiks, Uzbeks, Arabs, and Pakistanis, which play a critical and lethal role as trainers and advisors.[14] As important, Kilcullen notes the second tier of Taliban fighters are generally made up of local farmers that fight locally and

---

[11] Kilcullen, David. *The Accidental Guerrilla: Fighting Small Wars in the Midst of a Big One*. Oxford: Oxford UP, 2009. 70-114. Print.
[12] Ibid., 83.
[13] Ibid., 83-84.
[14] Ibid., 70-114.

"rarely on their own."[15] According to Kilcullen, this is modeled after the "focoist" strategy executed by the insurgent leaders such as Che Guevara and embraced by the Pakistani Special Forces doctrine, which relies on wandering bands of loyalists to rally the population against the government.[16] To this point, Kilcullen is remarkably insightful and should recognize the impact of cutting off these foreigners from the domestic population would have on the insurgency, but instead he falls short and fails to make the connection. This is not by accident, but by a deliberate discounting. In fact, Kilcullen splits an entire page worth of information dedicated to "external, active-sanctuary components of the Taliban" in Pakistan, but then states they don't require more attention because "they do not directly affect the situation inside Kunar."[17]

Kilcullen's statement seems to directly contradict the evidence he has presented. Further, Kilcullen concludes on a defeatist note stating in his view, "since there is very little practical prospect of the active sanctuary diminishing any time soon, the point is somewhat moot: the sanctuary's role in enabling the insurgency is a fact of life."[18] Instead of making what could be a very powerful connection to critical insurgent vulnerability, he shifts focus to tribal networks and the processes of political maneuver, which he elevates to a decisive element. This is curious as Kilcullen himself later writes that a region-wide approach will be key to denying the active sanctuary and that it will be essential to focus on disrupting safe havens, controlling borders,

[15] Ibid., 84.
[16] Ibid., 86.
[17] Ibid., 86-87.
[18] Ibid., 87.

undermining terrorist infrastructure while building a diplomatic environment hostile to terrorists and insurgents.[19] As such, Kilcullen is correct in recognizing something must be done about the sanctuary, but makes a serious error in suggesting an integrated strategy with Pakistan is the solution.[20]

Kilcullen fails to recognize it is not in Pakistan's inherent interests to remove this sanctuary and help defeat the Taliban anymore than it was in Pakistan's interests to help the Soviets defeat the Mujahedeen. In reality, the answer will need to come directly from the US/NATO force integrated with elements of the Afghan army and border police to effectively seal the border. Initially, this should primarily be a US/NATO effort, but over time, this should transition to a predominately Afghan force. Then, and only then, can one engage in an effective political strategy with Pakistan. Over time, this may ultimately yield the desired shift in Pakistani opinion toward the Taliban, which precipitates the Taliban's ultimate demise. However, waiting on a political solution will not occur within a timeframe that coincides with a victorious departure from Afghanistan by US/NATO forces.

US Military Doctrine

US Military doctrine provides next to no discussion on the subject of border security. The US military's cardinal guide to counterinsurgency, FM 3-24, *Counterinsurgency*, does mention border and sanctuary issues sporadically, but not as significant components of an overall strategy. The field manual also commits a

---

[19] Ibid., 111.
[20] Ibid., 111.

major error by failing to cite the advantages a counterinsurgent could gain by the removal of insurgent sanctuary. In fact, the word "border" appears only 22 times throughout the entire 282 page document, and this includes references to unrelated issues such as the NGO "Doctors Without Borders." Most of the references are indirect and are part of generic statements of what most COIN practitioners would consider obvious, such as: (1) Insurgents can derive benefits from sanctuary; (2) Contiguous borders with states providing sanctuary to insurgents are a vulnerability; (3) COIN should seek to isolate the insurgent from his base of support; and (4) The training of local national border police by US border and customs experts could be beneficial.

The most detailed example the document provides of the need for border security comes indirectly from a short section on securing the Iraq city of Tal Afar. It mentions the construction of a berm and other physical security measures to seal off the city from the Syrian border, and how these actions directly led to security within the city and the defeat of the insurgents. Nonetheless, the manual never gives proper credit to the effect of what could be termed "internal border security," or the securing of isolated population centers as opposed to conducting a linear external border security operation along a state's geographic national boundaries.

The former strategy became synonymous with the "Ink Blot" strategy in Iraq, where it was well suited to the terrain (large unpopulated desert with concentrated pockets of population), but was never understood from the perspective of border security. In fact, border security is at best given second or third tier importance throughout the document, and addressed in ways such as securing lines of communication (LOC). Although FM

3-24 does state the counterinsurgent should "make every effort to stop insurgents from bringing material support across international and territorial borders," it fails to explain when this is or is not important, how important it is, various ways and means to conduct border security, and most importantly, it never raises border security to the level of a necessary and critical part of a COIN strategy.

Border security is also markedly absent from other key US military manuals related to COIN operations. The US Army's FM 3-07, *Stability Operations,* does not even mention border security. FM 30-20-3 *Foreign Internal Defense Tactics, Techniques, and Procedures for Special Forces,* dedicates a total of one short paragraph, in a 219 page document. Its contribution to the subject can be condensed to the recommendation that local national forces should conduct border security, while their Special Forces advisors might need to assist in isolating insurgent forces.

No specifics are given on the importance of border security, when it is appropriate, or how to conduct such operations. The manual fails to discuss the problems associated with trusting locals to police their borders, when it is often those same locals who earn their livelihoods from cross-border smuggling, as is the case with the Pashtun tribes that straddle the Afghan-Pakistan border.[21]

Even more indicative of the lack of border security doctrine in the US military, is the 2004 Joint Publication

---

[21] Chandrasekaran, Rajiv. "Afghan Colonel Vital to U.S. despite Graft Allegations." *Washington Post - Politics, National, World & D.C. Area News and Headlines - Washingtonpost.com.* 4 Oct. 2010. Web. 16 Dec. 2010. <http://www.washingtonpost.com/wp-dyn/content/article/2010/10/03/AR2010100304094.html>.

3-07.1 *Joint Tactics, Techniques and Procedures for Foreign Internal Defense,* which is a 167 page document containing only two uses of the word "border," one of which is a passing reference to training personnel at ports of entry and exit, and the other is in a quote.

A dearth of doctrine on border security strategy is noticeable in older COIN doctrine as well. Foundational Marine Corps' doctrine was refined over years of fighting on Caribbean Islands such as Haiti and in Latin America, primarily in the late 1800's and early 1900's. These experiences produced the Marine Corps' *Small Wars* Manual, which does not directly address the issue of border security, and only provides two paragraphs on interdicting insurgent logistics.

This may in part be because islands were naturally isolated, and mainland wars were of a limited, though brutal nature, unconstrained by the current laws of war that restrict US COIN strategy. It could be argued that this freedom of action allowed fewer troops to pacify much larger areas through the use and threat of violence. This was evident in the successful suppression of the insurgency during the Philippines War (1899-1902) by a low number of troops relative to the Filipino population using what today would be considered brutal tactics and war crimes to pacify the islands.

An exhaustive review of military doctrine shows it was only during the Vietnam era that border security was addressed in any detail, although the lessons learned have since been forgotten or purged from military doctrine. FM 31-55 *Border Security/Anti-Infiltration Operations* contained the most detail directly related to border security, but even this field manual can only be accessed through a secondary source (*Out of Bounds: Transnational Sanctuary in Irregular Warfare;*

Thomas A. Bruscino, Jr.), which notes that only one copy of this FM was found and is only obtainable on microfiche at Ft Leavenworth's Combined Arms Research Library.

The US Army's FM 90-8, *Counterguerilla Operations*, contains four pages on border security, but also appears to have drawn primarily from the Vietnam experience, recommending forced relocation of civilians, which is not a viable option under today's limitations on warfare. Furthermore, the 1986 version of FM 90-8, *Counterguerrilla Operations*, recommends the US military should rely on indigenous forces to conduct border security operations.

This suggestion has proven to be an ineffective means of securing a border, particularly in Afghanistan. The remaining recommendations are so limited they can in no way be characterized as doctrine that stresses the importance of border security.

Case Studies

The following case studies will provide a chronological overview of the events of the Greek Civil War and the Algerian War of Independence. These case studies will recount the relevant information of the respective struggles as they related to state supported, cross-border insurgency. In particular, the case studies will pay close attention to the status of forces and relative success or failure of both the insurgents and the counterinsurgents leading up to, during, and after the implementation of border security. Each case study will conclude with an analysis of the findings. The case studies will seek to extrapolate, identify, and conclusively determine whether or not there is a correlation between effective border security and effective counterinsurgency.

Part 1

Lessons from History

Chapter One

The Greek Civil War

The Greek Civil War, fought from 1946-1949, provides a well-documented account of a state supported, cross-border insurgency in which physical border security was implemented. This conflict also presents a strong test of the thesis' hypothesis, since it is possible to compare the progress of both the insurgent and counterinsurgent, before and after the Yugoslavian border was closed in July 1949. This case study relies heavily on the seminal works written about this conflict by Edgar O'Ballance and C. M. Woodhouse, which are then augmented by other ancillary sources. Their works provide a number of primary sources written in Greek that are unavailable in translated form. These works provide immense insight on the events of the civil war.

The Greek Civil War was fought between the Greek Nationalist Army (GNA), backed by the United Kingdom and the United States, and the Democratic Army of Greece (Greek initials DSE), led by the military branch of the Greek Communist Party (Greek initials KKE), backed by Bulgaria, Yugoslavia and Albania. Although the political struggle had begun years before in 1943, after the German occupation of Greece during World War II, this case study will focus only on the last phase of the struggle from 1946-1949, which comprises the military phase of the struggle, and is generally recognized as the "Greek Civil War."

Geographically, Greece is a mountainous country roughly the size of the state of Alabama, with a mild, temperate climate of hot dry summers and wet winters (snow in the higher elevations). Today, Greece has a population of around 10.75 million people. It is a peninsular country possessing an archipelago of about 2,000 islands. At the time of the civil war, Greece was bordered in the north by Albania, Yugoslavia, Bulgaria, and Turkey. Of these, all except Turkey provided sanctuary and support to the communist insurgents. These neighboring sanctuaries created a front for cross-border operations roughly 635 miles long: 175 miles with Albania, 307 miles with Bulgaria, and 153 miles with Yugoslavia (See figure 1-1).[1]

Figure 1-1. Balkans.

[1] "CIA - The World Factbook, Greece." *Central Intelligence Agency*. 9 Nov. 2010. Web. 16 Dec. 2010. <https://www.cia.gov/library/publications/the-world-factbook/geos/gr.html>.

The Greek Civil War began slowly in 1946, after Soviet leader Joseph Stalin approved the formation of the Democratic Army (Greek communist guerrillas known as DSE) from Greek communist refugees residing in Yugoslavia.[2] This was preceded by the election victory of the internationally recognized Greek government in 1946, which the KKE boycotted and viewed as a catalyst for civil war. Although the Soviet Union gave the initial go ahead for the armed insurgency, it was the newly formed communist countries of Yugoslavia, Albania, and Bulgaria that provided the material support and sanctuary to the DSE forces throughout the struggle.

In the spring of 1946, the initially loosely organized bands of guerillas began to conduct cross-border raids against isolated villages, in order to establish a foothold in Macedonia and isolate it from the rest of Greece. The total insurgent strength by the summer of 1946 was estimated to be about 1,600 fighters operating in small bands of 5-20 men.[3] These raids provided combat experience, as well as food, supplies, and a source of new recruits and were targeted against weakly protected locations. The attacks had the result of forcing the Greek Gendarmerie to concentrate in more central locations for safety, which accelerated the situation as it left the smaller villages and towns vulnerable. The DSE raids were met with little resistance as the northern mountainous border of Greece was largely undefended and provided easy avenues for clandestine infiltration and escape.

---

[2] O'Ballance, Edgar. *The Greek Civil War, 1944-1949*. New York: Praeger, 1966. 122. Print.

[3] Woodhouse, C. M. *The Struggle for Greece, 1941-1949*. London: Hart-Davis, MacGibbon, 1976. 178. Print.

These early guerilla successes bolstered the morale and the confidence of the guerilla bands. In response, the Greek government at first classified the attacks as banditry and not part of an organized resistance, but by July of 1946, it became clear these bands of DSE insurgents were becoming more organized and were receiving foreign aid.[4] This prompted the government to question the adequacy of their policing operations and ability of National Guard garrisons to stem the increasing violence, but no immediate actions were taken. It wasn't until September 1946 that the government took the first real steps to quelling the insurgency, and placed the Gendarmerie under the Army's control.[5]

By August 1946, the KKE had selected Markos Vaphiadis (General Markos), to take charge and organize the DSE. In September, "General" Markos, arrived in Bulkes, a Yugoslavian army camp northwest of Belgrade to take command.[6] At the same time, Yugoslavian leader, Josip Tito and Bulgarian leader Georgi Dimitrov began taking a much more openly adversarial position towards Greece.[7] State officials from both countries also made numerous statements that made it explicitly clear they were at minimum supporting the insurgents ideologically,[8] and as such, both countries began to more openly allow the DSE sanctuary in their countries. This included allowing the DSE to increase their cross-border raids.

In addition, Yugoslavia moved forces to its border frontier with Greece to prevent the far weaker Greek

[4] Woodhouse, *The Struggle for Greece*, 185.
[5] Ibid., 187.
[6] O'Ballance, *The Greek Civil War*, 122.
[7] Ibid., 124-125.
[8] Woodhouse, *The Struggle for Greece*, 188-189.

military forces from attempting to cross the Yugoslavian border in pursuit of insurgents. The Greek military, aware of its inferior strength, had no choice but to stop short at its border in order to avoid risking interstate warfare with the combined communist forces of its neighbors to the north (Albania, Yugoslavia, and Bulgaria). This bolstered the confidence and emboldened the DSE to plan and stage ever more aggressive attacks.

The inability of the Gendarmerie and the Greek National Guard to check DSE attacks soon enabled General Markos to establish DSE bases throughout northern Greece by the fall of 1946. These dispersed bases inside Greece allowed the DSE to exert increasing control over the population and expand its intelligence network. This provided for further territorial gains south toward the city of Larissa in Thessaly. In response, the Greek government declared martial law over northern Greece. This was to no avail as the DSE continued to increase the scale of its attacks and increase its control of northern Greece. In October, the DSE was able to launch an attack on the town of Naoussa with a force of about 400 men aided by a now significant communist fifth column operating throughout much of Greece. This DSE force overwhelmed the National Guard detachment and ransacked the town for supplies before withdrawing.[9]

By November 1946, General Markos had expanded the DSE from around 4,000 fighters to a strength of more than 6,000[10] (estimates of total DSE strength vary, depending on how "insurgents" are categorized and whether fifth column supporters are included, but high

[9] *The Greek Civil War*, 126.
[10] Ibid., 128.

end estimates place total insurgent end year strength as high as 13,500 with an additional 12,000 in training camps outside Greece).[11] Further, attacks on towns grew more frequent with the DSE often holding the towns for days until the government could send reinforcements. The towns of Skra, Jannina, and Grevena were all sacked for supplies, government sympathizers were killed, and recruits were often forcibly conscripted into the growing guerilla ranks.

As a result, the Greek government, supported still by the British, recognized that the National Guard was simply not capable of dealing with what was now recognized as a civil war and authorized the use of the GNA. The GNA at the time consisted of about 100,000 ill equipped and poorly organized troops. These forces were deployed in a static posture as they possessed limited mobility. This allowed the DSE to easily outmaneuver the GNA positions and carry-on successful raids and limited seizures of small towns and villages throughout northern Greece. Even worse was the growing intent of the British to withdraw its troops and support as soon as possible leading to a desperate situation in which the Greek government sought international support from the United States.[12]

The entrance of the GNA into the conflict initially appeared to have little effect on DSE operations. By January of 1947, General Markos was in control of more than 100 Greek villages and had in excess of 7,000 fighters now in the DSE.[13] Further, limited military aid began to be provided primarily from Yugoslavia and to a lesser extent Albania. This included rations,

---

[11] *The Struggle for Greece*, 186.
[12] Ibid., 191-198.
[13] *The Greek Civil War*, 131.

ammunition, explosives, small arms, some mortars, some trucks, and training facilities. This allowed the DSE to begin transition to a more conventional, professional force during the winter while General Markos continued to expand his intelligence and logistics inside Greece. By the end of the winter, in March of 1947, the DSE emerged with 13,000 plus fighters and a headquarters permanently established in Greece, near Lake Prespa, at the intersection of the Greek, Albanian, and Yugoslavian borders.[14]

This area was specifically chosen for its close lines of communication with its cross-borders sanctuaries and the easily defendable nature of the Grammos and Vitsi mountain ranges. The GNA emerged frustrated and demoralized by their inability to stem the increasing power of the DSE. Often the GNA chased guerilla elements only to have to halt and watch the guerillas escape across a border into diplomatic sanctuary. Even more embarrassing were their claims to have cleared the Peloponnese and Lakonia of guerillas only to have 30 government forces killed two months later in the neighboring province of Messinia. This led the government to begin mounting conventional style military operations against the guerillas beginning in April and continuing throughout the summer.[15]

In a major attempt to stem the DSE, the Greek government reached out to the United States for assistance on March 3, 1947. Nine days later, US President Truman asked Congress for $400 million in aid to Greece.[16] Bolstered by this, in April, the GNA launched a division sized attack with about 15,000

---

[14] Ibid., 133.

[15] *The Struggle for Greece*, 204.

[16] *The Greek Civil War*, 137; *The Struggle for Greece*, 201.

troops against DSE strongholds in the Pindus, Khasia, and Agrapha mountain ranges.[17] The attack succeeded in surprising the DSE and reclaiming territory, but it was a temporary victory. The GNA failed to achieve a decisive victory because the DSE was able to withdraw intact across the borders into sanctuary. The GNA failure wasn't for lack of effort, but of understanding. The Greek forces attempted in vain to envelope the insurgents so they could not escape, but failed to take into account the insurgents were not bound to roads and thus, were able to extricate themselves from the government noose via remote mountain passes not accessible to vehicle traffic.[18]

Soon the DSE forces were re-infiltrating the area just cleared and had reestablished themselves by summer. The failure to achieve decisive results after much effort led General Tsakalotos, the commander of Operation Terminus (the overall summer offensive) to conclude in retrospect that government gains were insignificant.[19] Nonetheless, the American aid began to flow into Greece after President Truman signed a $300 million aid authorization bill on May 22nd, 1947. This aid included advisors to retrain the Greek army.

The DSE was clearly not hampered by the increasing American aid and stepped up military operations in 1947. In March, at the height of the first major GNA offensive, the DSE had about 13,000 fighters, but by May they had around 18,000, and by July they boasted a threatening 23,000 fighters. They also had up to an additional 8,000 fighters training in DSE camps located

[17] *The Greek Civil War*, 137.

[18] Woodhouse. 206-207.

[19] Woodhouse. 204.

throughout Yugoslavia and Albania.[20] Further, General Markos had established communist intelligence cells throughout Greece called YIAFAKA. YIAFAKA personnel probably had as many as 50,000 members at its height in 1947.[21]

Commensurate with the increase in personnel was an increased need for supplies. Yugoslavia, Albania, and Bulgaria increased their support to equip these forces and also began providing heavy weapons such as 75mm field artillery and anti-aircraft guns. This led to the first downing of a Greek Air Force plane on May 22nd, 1947 by the DSE.[22] Leveraging this increase in fighting power, General Markos, at the urging of Yugoslavia, began larger scale attacks against larger Greek towns. Beginning on May 28th, Markos employed a force of about 650 men unsuccessfully against the garrisoned town of Florina. He then proceeded to unsuccessfully attack Kilkis, Konitza, Kastoria, Grevena, and finally Alexandroupolis on July 31st.[23]

These battles inflicted a heavy toll in casualties on the DSE, shocked morale, and demonstrated that the DSE had not mastered cohesively fighting en masse. The battles also equally shocked the GNA by the DSE's boldness and large scale conventional nature as well as the fact it was nearly successful. As a result, by the middle of 1947, the Greek government had lost its control over large areas of the country and was in danger of real collapse.[24]

In response to the recognized DSE shortfalls in conventional style operations, the DSE's Balkan allies

---

[20] *The Greek Civil War*, 143.
[21] Ibid.
[22] Ibid.
[23] Ibid., 144-147.
[24] *The Struggle for Greece*, 208.

agreed to step up their technical support, providing items such as vehicles and communications equipment to allow for better large unit mobility, and command and control in battle. They also stepped up training of new recruits and established a four-week basic training program. At any given time between 3,000 and 7,000 new recruits were soon moving through this training pipeline. By mid-1947, the bulk of DSE supplies were now coming from outside Greece and it was estimated all of the mortars, anti-tank weapons, and artillery, as well as 75% of the DSE small arms, came from Balkan sources.[25]

This allowed the DSE to maintain a fighting strength of around 23,000, to return to the Pindus mountain region and solidify its hold during the summer.[26] This re-taking of territory by the DSE continued throughout the summer and fall leading to the GNA withdrawing from most of northwestern Greece. During October 1947 alone, the DSE raided over 83 villages.[27] This culminated in the declaration by the DSE of the "Free Democratic Greek Government" on December 24th, 1947. The next day, the DSE launched a large scale conventional assault on Konitza to establish a provisional capital.[28] However, although the DSE held out till January 7th, 1948, the GNA were ultimately successful in pushing them out of Konitza.[29] This battle was a watershed, as it marked the transition of the DSE from a primarily guerilla force to one primarily of a conventional organization using now both conventional and guerilla tactics. It also showed

---

[25] *The Greek Civil War*, 153.
[26] Ibid., 150-151.
[27] Ibid., 156.
[28] Ibid., 158.
[29] Ibid.

that after a year of hard fighting, the GNA had failed to check the growing power of the DSE.

1948 is often referred to as the year of the stalemate. Both the DSE and the GNA increased the size and the competencies of their forces during the year, but neither side was able to deliver a decisive defeat. On the ground, the lack of victory for the GNA meant the Greek government was losing. Further, the Greek government estimated that between March 1947 and March 1948, more than 9,000 insurgents were killed with that number again wounded or captured, but the DSE still continued to carry out attacks and increase in strength showing a remarkable resilience. By March 1948, the DSE had hit a high of 26,000 fighters.[30] Then on May 10, 1948 the DSE delivered another strong message to the GNA that it had continued to improve its weaponry after shooting down two of its planes.[31]

Even more remarkable is that the DSE continued to check the GNA even as the GNA received massive aid from the United States. By May of 1948, it had increased its military strength to 168,000 personnel in the navy, air force, and army plus another 50,000 National Guards.[32]

In May of 1948, the government claimed to have largely cleared large parts of Macedonia and Thrace of rebels.[33] However, like in past operations, the insurgent fighters simply withdrew across the Yugoslav and Bulgarian borders to safety only to reemerge soon after to launch attacks and reoccupy territory in Greece.[34] This was verified in numerous reports by the United

[30] Ibid., 165.
[31] Ibid., 163.
[32] Ibid., 166-167.
[33] *The Struggle for Greece*, 240.
[34] Ibid., 240.

Nations Special Committee on the Balkans (UNSCOB) that explicitly stated the DSE had "received aid and assistance from Albania, Bulgaria, and Yugoslavia."[35]

These reports went on to note examples of convoys crossing the borders, heavy supporting fires originating in Albania, and prisoner statements claiming to have been trained and equipped outside of Greece. On June 19th, the GNA launched a force of 40,000 against about 8,000 insurgents to clear them from the Grammos Mountains. DSE total strength had finally peaked at around 28,000, but only a total of 12,000 could be concentrated to counter the GNA offensive.[36]

Even against these staggering odds, the insurgent forces held the government forces, which were backed by heavy artillery and recently delivered American aircraft. It was not until August 20th, 1948 when the DSE conducted a masterful fighting withdrawal into Albania and sanctuary where they were able to regroup and infiltrate back into the Vitsi Range just north of where they were pressed out.[37] The GNA then pressed a late August offensive against the DSE in the Vitsi Range, but was repulsed by heavy DSE resistance. These operations continued throughout September against the DSE until both sides finally wore down into a general stalemate and only smaller scale engagements and DSE raids were carried out. It is important to note that at this point in the civil war the DSE and KKE (communists) did not have the support of the majority of the population, but nonetheless were still able to successfully fight major conventional battles against the government forces.[38] To

[35] *UNSCOB Report, General Assembly*. Rep. Vol. A/574. New York, 1948. Print. Para 188.
[36] *The Greek Civil War*, 170-171.
[37] Ibid., 173-174.
[38] Ibid., 175.

this point, General Tsakalotos stated that "the balance had begun to turn against the national forces."[39]

By the end of 1948, it was clear the Greek government despite staggering advantages in numbers, equipment, and training were unable to decisively defeat the DSE, and in fact, had been beaten across all fronts. A force of no more than 13,000 insurgents had effectively repulsed a government force of more than 50,000. Further, although the DSE suffered staggering losses of greater than 32,000 casualties for the year, General Markos managed to maintain an effective fighting force of just about 21,000-25,000 with high morale.[40]

The civil war ended in 1949. The events that occurred in the ensuing months fundamentally affected the DSE and led to a precipitous and rapid deterioration of the DSE's ability to sustain resistance. This change in the operational environment ultimately led to the GNA inflicting a decisive defeat on the DSE. By October, a ceasefire was declared and the war was over. The events that led to this defeat after years of sustained conflict are perhaps the most telling for the interests of this case study.

Three key events occurred during 1949 that set it apart in the war and precipitated the final conclusion of hostilities. First and most important, a political rift that had been developing for some time within the communist camp cemented a rift between Stalin and Tito. This ultimately led to the closing of the Yugoslav and then the Albanian border. The second event was that the GNA began a new campaign of internal border security starting from southern Greece of systematically

---

39 *The Struggle for Greece*, 243; Tsakalotos, Thrasyvoulos. *Khrónia Stratiótis Tis Elládos*. Rep. Vol. Th: 40. Athens, 1960. Print. 165-169.

40 *The Greek Civil War*, 176-178; *The Struggle for Greece*, 257.

clearing and separating the population from the insurgents. This culminated with the remaining DSE forces being split and forced across a border where they were disarmed and prevented from returning to the fight. Third and finally, the DSE chose to stand and fight conventional set piece battles against a numerically superior and better armed enemy.

Beginning in November of 1948, the KKE decided it must make a major political shift to support either Stalin or Tito. The implications were dire and no situation presented a good scenario. If the KKE supported Tito, it would lose Albanian and Bulgarian support, which remained in the sphere of Stalin. If the KKE supported Stalin, it would lose Yugoslavia's support and risk losing its middle, which would cut the DSE off from its units in the east and the west. Further, a power struggle developed between the strategies of General Markos and Nicholas Zakhariadis. Markos firmly believed that the DSE should continue conventionally organized guerilla resistance to wear down and ultimately defeat Greece. However, Zakhariadis believed that the DSE must move into fully conventional operations to deal a decisive blow against the GNA while they seemingly had the advantage and before the loss of sanctuary from the split in the communist party could become relevant. By 1949 these fateful issues were decided and the KKE would side with Stalin while General Markos would be replaced by Zakhariadis.

The DSE wasted no time and by December 1948, were conducting large scale assaults against towns. After learning difficult lessons, the DSE refined its tactics and by January 12, 1949 successfully took over the town of Naoussa.[41] The DSE repeated victory and soon took the

[41] *The Greek Civil War*, 183.

town of Karpenisi in late January. It wasn't until late March that the GNA was able to dislodge the communist forces. However, cracks were appearing in the DSE. General Papagos took over command of the GNA and immediately set off on a new clearance strategy. This strategy included massive relocation of civilians to secure areas free of insurgents beginning in the south and moving north. As the GNA tightened the cordon, enhanced by the geography of the southern peninsular region, the DSE was left with nowhere to hide, resupply, or run.

This effectively isolated the DSE from the population via internal border security. By March 16th, General Papagos was able to announce that he had cleared the southern Peloponnese of insurgents. Concurrently, in February, Zakhariadis launched a massive assault on the town of Florina. This was to be his first major victory, but in short order, the attack collapsed and it became his first major set-back at the helm of the DSE. Further, the first major impact of the communist split was being felt as Tito began a decrease in aid and implementation of restrictions on cross-border movement. On May 23rd, UNSCOB confirmed this and reported a significant decrease in Yugoslav aid moving across the border.[42] This forced the DSE to move its primary base of operations to Albania and Bulgaria.

Still strong and resolute, the DSE reemerged from Albania in April armed with the latest Soviet weapons and equipment and quickly recaptured key terrain in the Grammos Range before the GNA could respond. These forces were also able to seize towns in eastern Macedonia and Thrace.[43] However, the DSE was no

---

[42] *The Struggle for Greece*, 272.

[43] *The Greek Civil War*, 194.

longer able to recoup losses as recruiting grounds in southern Greece were now internally sealed off by the GNA. Even worse, on June 21st, DSE resistance was broken in central Greece nearly eliminating the DSE's ability to recruit as well as effectively blinding its intelligence network. This meant that losses by the spring of 1949 had reduced the DSE to a still formidable 19,000 fighters.[44]

July marked the beginning of the final unraveling of the DSE. On the 10th, Tito announced he would close down his borders, which was completed in amazing speed. By July 21st, UNSCOB reported the Yugoslav border had been effectively closed and further confirmed by Kardelj, the Yugoslav Foreign Minister.[45] This was the disaster the DSE feared as their forces would now be split and isolated in the west on the Albanian border and in the east on the Bulgarian border.[46] To offset this, aid from Bulgaria was increased to the DSE. July also marked the beginning of another GNA offensive aimed at DSE positions on the Yugoslav border.

After heavy fighting, the insurgents were forced to retreat across the border into Yugoslavia. However, unlike before, the DSE forces were disarmed and prevented from returning into Greece. Now with the DSE split, the GNA focused on the DSE stronghold remaining in the east backed Albania. Here the DSE had created fortified fixed fighting positions manned with about 15,000 fighters intending to hold out at all costs. The GNA responded in kind by building a physical barrier to prevent any DSE fighters from infiltrating

---

44 Ibid., 192.

45 *The Struggle for Greece*, 273.

46 *The Greek Civil War*, 195.

back into central Greece and effectively sealing the DSE against the northern border.[47]

The GNA's offensive began on August 5th, and although encountering stiff resistance, the massive firepower (that included new American Helldiver aircraft capable of pin point attacks with heavy ordnance) finally overwhelmed the fixed DSE positions by August 16th.[48] Upon defeat, the DSE followed previous doctrine and withdrew across the borders into Albania and Yugoslavia. The forces that withdrew into Yugoslavia, over 1,000, were disarmed by the Yugoslav forces and not allowed to return to the fight. The rest, about 6,000, reformed in Albania and reinforced the remaining Grammos pocket of resistance with about 2,500 fighters for a total fighting strength around 8,000.[49]

Heavy fighting between DSE pockets of resistance and the GNA continued throughout August, but ultimately, in set piece battles, the superior firepower and numbers of the American supported GNA wore down and forced the withdrawal of DSE forces across the northern border. The insurgents that entered Yugoslavia were disarmed and detained, but the insurgents managing to make it to Bulgaria and Albania still had the means to resist. However, this changed on August 26th, 1949 when Albanian General Hoxha decided that he could no longer risk provoking the now well-equipped and highly competent Greek military into a conflict without Yugoslavia's backing, and so, announced that all armed

---

47 Ibid., 196.
48 Ibid., 197.
49 *The Greek Civil War*, 281-282.

Greeks found in Albanian would be disarmed and detained.[50]

This effectively removed Albanian sanctuary and closed its borders. Now Zakhariadis had only Bulgaria to rely on for sanctuary and desperately tried to reform his forces there to carry out a guerilla struggle. This was not to be. By August 30th, Operation Torch, the last major GNA offensive, successfully eliminated the remaining resistance in the Grammos Range sending the remaining DSE fighters retreating into Albania. Stalin ordered a ceasefire and on October 16th, 1949, the DSE Radio announced the ceasefire. Although 3,500 were thought to have made it to Bulgaria with maybe another 1,000 still hiding out in Greece, the DSE no longer had the means to pursue resistance and the war was over.[51]

Analysis of the Greek Civil War

The Greek Civil War provides the opportunity to examine how effective border security and removal of sanctuary affected a powerful communist insurgency that had raged for almost four years and at one point witnessed the insurgents in control of nearly 4/5 of Greece and the Greek government near collapse. It was clear that throughout the war, the government forces continued to become better trained, equipped, and led, as well as nearly doubling their forces, but at no time before Greece's borders were closed could the government forces effectively deal the insurgents a decisive blow. Again and again, the GNA launched large scale offensives clearing large areas of Greece, but failed to prevent their reconstitution and return to these areas.

---

[50] *The Greek Civil War,* 200.
[51] Ibid., 202.

The resilience of the insurgents even after suffering tens of thousands of casualties appears as a testament that body counts alone are not symptomatic of progress or victory.

Considering the above, the advantages provided by the state supported cross-border sanctuary to the communist insurgents have to be identified, not as a critical vulnerability, but as the actual center of gravity of the insurgency. This refocused assessment of the enemy is warranted as it appears no other government tactic, technique, or procedure appears to have had this kind of catastrophic impact on the insurgency. In fact, massive clearance operations, police sweeps, prosecutions, amnesty periods, heavy direct combat, and political maneuvering with international bodies such as the United Nations came to naught when compared with the swift effects resulting from the removal of Yugoslav and Albanian support.

Thus, the cross-border sanctuary was essential in preventing the insurgent forces from being cornered and destroyed in conventional combat as they repeatedly were able to conduct tactical withdrawals across the Greek border into sanctuary only to reemerge at the nationalist army's rear and counter-attack to much success. This ability to continually outmaneuver the GNA and escape to sanctuary was as effective for the insurgents as it was frustrating for the counterinsurgents. This sanctuary was also a conduit for training and large supplies of weapons, to include artillery and anti-aircraft guns, which simply was beyond the reach of Greek forces. However, before one can conclude that the closing of the border and removal of state supported sanctuary was the critical factor leading to the defeat of what to that point appeared to be an insurgent force near victory, counter arguments must be addressed.

First, one may argue that is wasn't the closing of the borders, but the DSE's decision to conventionally fight that actually led to its swift defeat in the middle of 1949. This critique does not hold up under close scrutiny. In fact, the insurgents had been effectively operating in a conventional fashion since 1947, long before they peaked in power. Further, the DSE achieved numerous victories operating more or less conventionally and affected a stalemate and then reversal of conventional government forces by the end of 1948. Finally, even though the DSE engaged in large scale conventional style attacks and battles, it still retained tactical cross-border withdrawal as a key tactic throughout the entirety of the war irrespective of whether it was operating conventionally or unconventionally.

Second, one may argue that the military reforms and strategy of General Papagos actually led to the defeat of the insurgents in 1949. This claim also fails to be defendable. Although General Papagos did successfully conduct clearance operations from south to north in 1949, which removed the insurgents 5th column operating amongst the population and providing supplies and intelligence, this wasn't new. The government had conducted similar operations against the insurgents multiple times that included sweeping insurgents from the Peloponnese and large areas of Macedonia and the border regions. Each time the government announced a region as pacified, it soon found the insurgents again operating in strength within a few weeks to a couple months. Thus, although his tactics can be persuasively argued as more effective and efficient, they can't be said to be the conclusive element that brought about insurgent defeat.

Third, one may argue that it wasn't necessarily new tactics, but better organization and equipping of the Greek military largely by the United States and the provision of US advisors and training that finally overwhelmed the insurgents. Again, although the GNA showed marked improvements in battle and clearly benefited from the better training and weapons, this was not decisive. In fact, insurgents were able to demonstrate from at least 1948, that they were able to mitigate the improved firepower, larger numbers, and better training of the GNA, by countering with their own heavy weapons, improved equipment, and continued training and adaption.

Most importantly, the DSE effectively adjusted their tactics to include surprise attacks followed by planned tactical withdrawals before the government forces could bring to bear their heavy weapons. Further, it was only by choice that the insurgents stood to fight when they believed they had the advantage and always maintained the option to withdraw if the fight became unsustainable. This effectively removed the ability of this firepower to be decisive.

Fourth, one may claim it was actually the political split in the communist party that signaled the death of the insurgency. The argument would suggest the lack of cohesion inside the KKE and DSE led to poor decisions and collapse. However, the rift itself seems to have had little effect and actually appeared in 1948, when the DSE was still increasing in power and carrying out even more sophisticated attacks. Further, there appears to have been no major dissent in the ranks when command was shifted from General Markos to Nicholas Zakhariadis.

Although this rift later led to the closure of borders, which was the key factor, the rift itself was not sufficient to decisively doom the insurgency.

Fifth, one may argue that the insurgency was unable to repopulate its ranks by 1949, due to casualties from heavy fighting and the securing of the interior of Greece. This argument has merit in that the insurgency did suffer heavy losses, but it was far from combat ineffective entering the summer of 1949. The DSE was still fielding nearly as many or more fighters than it did in late 1947, it was better trained, and had much better equipment and weapons.

Further, nearly the same number of DSE forces (~6,000) withdrew to Albania in 1948 after blunting a major GNA offensive to clear the Grammos region and still effectively reemerged only a few weeks later slightly to the north in the Vitsi region to retake this area.[52] As a result, there was in fact, even in 1949, little expectation amongst government forces that the insurgents would not reemerge after being forced out of the very same areas during the summer of 1949 with what had previously proven to still be effective numbers for offensive operations against the GNA. As such, numbers alone cannot be the decisive element that led to the insurgent defeat.

Sixth, one may claim that the rugged mountainous border of northern Greece could not be effectively secured and thus, could not be reasonably argued as the decisive factor. Although this may defy contemporary thinking, Yugoslavia demonstrated that in less than two weeks, it could be effectively sealed off as testified by UNSCOB observers. Further, Albania followed suit shortly thereafter and also was effective in interdicting

---

[52] Woodhouse. 242.

and disarming insurgents entering its borders. Thus, one must conclude that even in extremely rugged mountainous terrain, when the will is present, governments can effectively secure their borders.

Finally, one might argue it was the removal of sanctuary and state support more than the border security, which was the most important issue. This argument follows RAND research that has identified a correlation between sanctuary and insurgent success (or conversely counterinsurgent failure). However, although this thesis argues the issues are inextricably linked to effective counterinsurgent strategy, in the case of the Greek Civil War, it was in fact the securing of the borders that proved more decisive.

Had Yugoslavia and Albania only stopped providing support, but allowed the insurgents to freely move across their borders, the war would have raged on indefinitely at least through guerilla type action because Bulgaria was still providing the training bases and military equipment needed to continue effective resistance and the insurgents still enjoyed the critical freedom of movement.

More specifically, Bulgaria and Rumania continued to provide both sanctuary and support to the insurgents, which would have allowed the DSE to continually maneuver its forces across the entire northern front with impunity and continue to arm and train with the latest Soviet weapons and equipment. Thus, much more devastating was the inability for the insurgents to move freely across the borders, which meant either the insurgents would ultimately be surrounded and destroyed inside Greece or forced out of Greece permanently.

Only the fundamentally altering of the political calculus of the nations supporting the DSE, by what in the case of Albania turned out to be a fear of Greek invasion and in Yugoslavia the rift with Stalin, did the Greeks enjoy a fortuitous change of events that resulted in both the borders being closed and state support being removed. However, had this situation not materialized, it is fair to conclude that only by unilateral Greek initiative, that affected the DSE in a similar manner through physically securing the border, could the GNA have achieved a similar decisive victory and cessation of hostilities.

The decisive action that ended the Greek Civil War was the closing of the borders with Albania and Yugoslavia magnified by the removal of state support and sanctuary. This single action, which denied the insurgent force the ability to move freely from engagement to cross-border safe haven proved more effective in stopping the insurgency than almost four years of heavy fighting and bloodshed. No other single action proved to be as decisive. From the time the borders were secured to the time the insurgency collapsed, barely two months had passed.

This event was such a powerful force that even without the securing of the Bulgarian border, the insurgency still collapsed. Arguably, the peninsular geography of Greece facilitated this by concentrating the border areas in the north while offering no means of escape along the coastal regions. This is why the Greek government enjoyed much greater success in pacifying the southern and central regions as the Navy was able to provide an effective external cordon allowing the GNA to effectively corner and eliminate any pockets of resistance.

The Greek Civil War offers the counterinsurgent fighting in Afghanistan reasons for both optimism and pessimism. On the optimistic note, the Greek Civil War definitely demonstrates a powerful insurgent force that has operated effectively with state support for years can be quickly and decisively defeated. It further shows although mountains can be a great asset for insurgent operations, they still can be effectively cleared and secured. This dismisses claims that the mountains are impregnable for counterinsurgent forces and are not conducive for border security strategies.

If Pakistan can be made to change its strategic calculus, and deny state support to the Taliban insurgents operating from within its borders and this is coupled with a deliberate border security initiative, a decisive victory may still be achieved. However, much more ominous are the hard lessons for Afghanistan. Even after suffering heavy casualties and being repeatedly swept from a country that is generally in control of an effective government and hostile to the insurgents, the insurgents managed to regenerate their forces and again return to threaten and control large areas of Greece.

Afghanistan has neither an effective government nor an overly adversarial population to the insurgent. In fact, it is the counterinsurgent that is looked at with hostility in much of the regions inhabited by Pashtuns. Also, the security force assistance (SFA) provided by the US in massive quantities that included training, advisors, and equipment also did not have a decisive effect, which matches similar experiences of the US in Vietnam and the Soviets in Afghanistan.

Further, even with massive firepower and numerical advantages, the GNA could not decisively defeat the

DSE. This suggests that NATO's dominance in firepower and technology will likely prove inadequate in and of itself to defeat the insurgency. Drawing in anecdotal lessons from the failed Soviet War in Afghanistan, where massive indiscriminate firepower was used, seems to support this conclusion.

As such, not until the border was secured, could the GNA bring to bear its firepower and drive home successes for a decisive result. Also ominous is the total lack of Pakistani will and/or ability to cease support for the Taliban and effectively seal its borders with Afghanistan. Ideally, if options are available that can move Pakistan to take decisive actions, this is ideal, but under the circumstances, the US/NATO force must recognize the need to unilaterally conduct this operation.

If the political calculus of Yugoslavia and Albania had not changed, it is unlikely the Greek Civil War would have ended in the manner it did and in the least, it certainly would have carried on for years without unilateral Greek action to secure the border. What this means for Afghanistan is that until the border with Pakistan is effectively physically secured by someone (be it Afghans, NATO, or the Pakistanis), the insurgent force will continue to avoid decisive engagement and most likely outlast even the most determined operations of NATO and the Afghan forces.

Chapter Two

The Algerian War of Independence

The Algerian War of Independence was selected as the second case study not only because it meets the criteria for a state supported, cross border insurgency that incorporated border security, but it also involves strong similarities to the war in Afghanistan in that it has similar long desert and mountainous borders, it involved a western counterinsurgent against an Islamic insurgent, it was transnational in nature, and it sustained for almost a decade before conclusion. Upon deeper review, one also finds many cultural similarities that have helped insurgents through adherence to tribal codes mandating principles such as hospitality to "all" visitors, asylum even to enemies, and bravery in warfare (known as Pashtunwali in Afghanistan and Pakistan).[1]

Further improving the comparison is the nearly identical length and similar topography of the desert/mountain Afghan border with Pakistan (2,430 kilometers) as compared to 2,524 kilometers of desert/mountain border Algeria shares with Tunisia and Morocco (See figure 2-1). Finally, and perhaps most important for this case study, is that border security was not initially implemented by the French. This provides a key chance to test the hypothesis by evaluating the before and after effects, if any, the physical barrier provided. The data and events used for case study evaluation are primarily constructed from the seminal works written on the war by Edgar O'Ballance and

---

[1] Horne, Alistair. *A Savage War of Peace: Algeria, 1954-1962*. New York: Viking, 1978. 88. Print.

Alistair Horne augmented by numerous external sources.

Figure 2-1. Algeria.

The Algerian War of Independence, fought from 1954-1962, was waged primarily between France and Algerian insurgent organizations, primarily the National Liberation Front (FLN), seeking independence from French colonial rule. The FLN received both sanctuary and state support primarily from Tunisia and Morocco, but also had strong transnational support and received diplomatic as well as materiel support from countries such as Egypt, Czechoslovakia, and even anti-colonial and pro-Islamic factions in Western Europe. However, the war was not neatly split between two belligerent factions as numerous other Algerian insurgent groups simultaneously fought each other generally aligned along lines supportive and against remaining a part of France.

The Algerian National Movement (MNA) was one such pro-independence faction that fought with the FLN. The French were no less split and also aligned by their position on Algerian independence, which led to at least two attempted coups and protests in Paris. The war embodied brutalities of all types by all sides, but by 1960, the French military had generally reestablished order. However, the war and its brutality led the French and Algerians to ultimately conclude that Algeria was to be independent. Shortly thereafter, French President Charles de Gaulle put the independence issue to a vote that paved the way for cessation of hostilities and full Algerian independence. The war ended officially in March 1962 with the signing of the Evian Accords, which granted Algeria its independence.

Algeria is a country approximately 3.5 times the size of Texas and is the second largest country in Africa. It is situated with its northern coastline along the southern coast of the Mediterranean Sea and it shares borders from west to east with Morocco, Western Sahara, Mauritania, Mali, Niger, Libya, and Tunisia. The climate is arid to semiarid with mild wet winters and hot dry summers along the coastline. In the mountain and plateau regions it is drier with colder winter. The southern sirocco is extremely arid marked by classic desert conditions.[2] Today it is the home to an estimated 34,586,184 people.

---

[2] "CIA - The World Factbook, Algeria." *Central Intelligence Agency.* 9 Nov. 2010. Web. 16 Dec. 2010. <https://www.cia.gov/library/publications/the-world-factbook/geos/ag.html>.

Chapter Two: The Algerian War of Independence

The lead up to the outbreak of hostilities in Algeria may have been gradual, but on the night of October 31st, 1954, a surprise offensive of at least 70 attacks and bombings across the country made for the clear beginning of a new and violent phase of resistance.[3] This attack was to be carried out through November 1st, All Saints Day, which correctly assumed most French officials in Algeria would be at home celebrating the holiday. As with the initial stages of many insurgencies, the population (including Muslims) and the French authorities viewed this as banditry or a small tribal revolt and did not initially understand the seriousness of what had just begun.

This was in fact the coordinated efforts of what became known internationally as the National Liberation Front or the FLN (in Algeria they were known officially as the Armée de Libération Nationalé or ALN), which at the time probably only could muster 2,000-3,000 fighters.[4] From this point on, the insurgents began conducting hit-and-run guerilla type operations that gradually became more coordinated and competent as the ranks began to grow and experience in fighting was gained. Upon the outbreak of hostilities the ALN had an estimated 350-400 firearms and no machine guns, so it intentionally avoided direct confrontations.[5]

Thus, these initial raids often focused on stealing and acquiring arms during the beginning period of the war. Like Mao's Chinese and other communist revolutions, the ALN modeled their tactics on first leveraging guerilla tactics to wear down the government forces and then

---

3 O'Ballance, Edgar. *The Algerian Insurrection, 1954-1962.* Hamden, CT: Archon, 1967. 39. Print.

4 Ibid., 39-42.

5 Horne. 84.

building into a conventional force. However, the ALN incorrectly assumed the French would quickly fall as other colonial regimes. This attitude of overconfidence was bolstered by the recent defeat and withdrawal of French forces from Indochina, known today as Vietnam, after the Viet-Minh decisively defeated the French forces garrisoned at Dien Bien Phu. Thus, began a long and bloody seven and a half year conflict that brought an end to the French Fourth Republic.

The French responded with a poorly conceived heavy handed approach, including mass roundups of many locals that were at least initially pro-France, but after harsh treatment, began turning more and more to the ALN. Further, French military actions to include patrols in the mountains began in a clumsy fashion. Quickly the French learned local insurgent intelligence was always one-step ahead of them, which prevented cornering insurgents conducting limited raids against their patrols and convoys and who then disappeared into the mountains. The French also were unable to protect the sympathetic locals that were soon paralyzed with fear by the targeted killings of the ALN.

Today's veterans may also find the ALN's use of mountaintop bonfires and dogs to warn of approaching patrols reminiscent of their experiences in the mountains of Afghanistan. It wasn't until French paratroop units arrived from the 25th Airborne Division, led by Colonel Ducournau, a seasoned counterinsurgent expert from the Indo-China war that the French began to improve their tactics and achieve some success.[6] With this came killings and arrests of a number of ALN leaders and forces pushing the insurgent's existence to the brink.

---

[6] Horne. 102-103.

One of the first competent ALN regional commanders to emerge was Ben Boulaid. He was in charge of Wilaya I, which was the insurgent name for its regional organizational structure that broke Algeria into six regional commands. He organized a more professional organization out of the mountains that sustained its growth by organizing the provision of smuggled weapons from Egypt via Libya until his capture in Tunisia by French forces.[7] During that time, they carried out limited raids and ambushes that usually resulted in the insurgents fleeing after decidedly poor results. Most other Wilayas followed a similar pattern of education through trial and error and by the summer of 1955, they began to emerge as coherent insurgent groups capable of organized resistance.

1955 also marked a tragic chain of events that turned the war to personal and total. On August 20th, the commanders of Wilaya 2 launched a brutal attack against French Algerians of European descent to polarize the war and force a heavy handed French response.[8] This attack killed at least 123 people, of which, 71 were European, in the most barbaric of manner.[9] Men, women, children, and babies were massacred. Babies were dashed against walls and mothers were savagely dismembered with their children. Men had their throats slit in their own beds and were gunned down in the streets.[10]

The savageness of this succeeded in polarizing the sides and the French response was swift and brutal. Paratroopers were sent in and although the final death

---

[7] O'Ballance. 44.
[8] Horne. 118-119.
[9] Ibid., 120-122.
[10] Ibid., 120-122.

counts will never be known, it was claimed that as many as 12,000 insurgents were killed.[11] This horrendous event became known as the Philippeville massacres and succeeded in bringing a flood of new recruits into the ALN. Nonetheless, it was the Muslims that suffered the worst, mainly by the hands of the ALN, and not the French, through internecine violence, pacification, intimidation, and internal power struggles against MNA sympathizers and supporters.

This occurrence of internal violence is actually common and has been witnessed repeatedly in insurgencies such as the partisan wars in Yugoslavia from 1941-1945 where Tito expended a great deal of his energy fighting Serbian factions.[12] This also has contemporary parallels amongst the fighting in Afghanistan between rival warlords, ethnic, tribal, criminal, and political factions. This internal sectarian violence perhaps could have been exploited by the French to eliminate the ALN, but opportunities were missed, and soon the door closed forcing a long and protracted struggle.

By the winter of 1956, the Wilayas were first able to begin to acquit themselves respectably against French forces. Wilaya V in particular managed to fight a stand-up battle against French troops and only a month later nearly besieged the city of Tlemcen.[13] This was illustrative of a growing competency across the ALN, but the insurgents were plagued by shortages in arms and supplies, which prevented an initial sizable growth in their force numbers and capability.

---

[11] Ibid., 102-103.

[12] Ibid., 135.

[13] O'Ballance. 48.

It wasn't until Egyptian President Nasser began moving arms from Czechoslovakia to Algeria in September of 1955, that the ALN was able to arm regular forces. As a result, it was April 1956 when the ALN's forces were finally estimated to have grown to around 8,500 fighters.[14] To combat this growing threat, the French forces launched sweeps into the mountains and established posts throughout where patrols and cordon-and-search operations could be launched from. However, the French were undermanned to control both the urban and rural mountainous regions, so these conventional operations proved to be of little value, since the insurgents easily bypassed the pockets of activity.[15]

As such, new areas of violence began to appear all over the country perhaps emboldened by the survival of the ALN throughout the winter and the fresh supplies from Egypt.[16] It is also worth noting that the French for some time believed that social and economic reasons were the root cause of insurgency and set about building hundreds of schools, medical centers, and implementing agrarian reforms all to little effect.[17] Instead, the French had more success in their operations after instituting a strategy of "quadrillage" where areas, particularly urban ones, were sectioned off and controlled by a grid pattern. Part of this strategy also included the resettlement of the population near patrol bases to isolate them from the insurgents and to leverage locals in the security. This was manpower intensive, but an increase in French forces eventually to 400,000 allowed for this to be more

---

[14] Ibid., 49-50.
[15] Ibid., 51.
[16] Horne. 111-112.
[17] O'Ballance. 53.

effectively implemented.[18] The result was by the spring of 1956, ALN units were again being hit hard by the French.[19]

The spring and summer of 1956 showed marked success in the French situation, but actions occurring simultaneously to Algeria's east and west would prove to be ominous for the French counterinsurgents. In March 1956, both Morocco and Tunisia were granted independence by France. Both of these new governments sympathized with the FLN and made immediate moves to not only establish direct routes for supplying the insurgents with weapons and equipment, but also established their territory as secure cross-border sanctuary for the FLN.[20]

By the fall, the French had failed to follow-up on their successes and gradually were losing the initiative just as the ALN was organizing itself in a much more cohesive entity with better command and control and clear lines of command. Fall of 1956 marked an escalation of terrorist violence in the city of Algiers as it descended into chaos and violence, which forced the deployment of the French 10th Paratroop Division commanded by General Massu to lock down the city and specifically the Muslim section known as the Casbah.[21]

General Massu took command of the operation in January 1957 and aggressively pursued the insurgents through a ruthless web of informants, surveillance, and direct action that led the ALN to flee for the first time to sanctuary in Tunisia or risk being completely destroyed.[22] This was a major blow to the ALN and a

---

[18] Horne. 113.
[19] O'Ballance. 64-65.
[20] Ibid., 67.
[21] Ibid., 80.
[22] O'Ballance. 80-81.

victory for French counterinsurgency effort, or so the French thought. In reality, it moved the ALN into the realm of a state supported, cross-border insurgency that was receiving transnational support. This sanctuary in Tunisia provided the needed breathing room for the insurgency to regroup, train, arm, and launch a new wave of more violent and effective attacks inside Algeria.[23]

Throughout 1957, the ALN grew in strength and numbers and began engaging in more conventional large scale operations. The most notable example in 1957 was the continued Battle of Algiers. It wasn't until the February 19th take down of a long sought bomb factory in the Casbah and seizure of a large supply of explosives, bomb making material, and intelligence, that resistance in Algiers unraveled.[24] By March, order was restored to Algiers, but at the cost of any remaining moral high ground. The French reliance on torture, which alienated the population was arguably necessary to win the battle, but has been correctly pointed out to also have lost the war.[25] These battles took a heavy toll on the French forces, but benefited the French in that they were able to leverage their superior conventional forces and inflict heavy casualties on the ALN allowing them to rightfully claim their first real major successes against the ALN.[26]

The French use of the quadrillage system still bore benefits and heavily suppressed insurgent activity, but it became less and less effective as the primary origin of insurgent attacks moved outside of the Algeria's borders throughout the year. The result of this was that it became

---

[23] Horne. 219.
[24] Ibid., 193.
[25] Ibid., 205-207.
[26] O'Ballance. 90-91.

apparent the static nature of the French forces tied down an exorbitant number of soldiers that could be better employed in a mobile fashion.[27] By the end of 1957, the ALN had 30,000 troops in Tunisia and the Tunisian government was so heavily involved, it tasked the National Guard to ferry weapons to the Algerian border where they could be picked up by the ALN and was overtly treating wounded fighters in its hospitals.[28]

On Algeria's western border, the ALN set up similar bases of operation in Morocco that quickly became problematic. This massive amount of better trained and well-armed manpower posed a serious and escalating threat to counterinsurgent efforts. To stem this growing tide, the French did achieve some success politically getting the Libyans to curtail the movement of arms through its territory from Egypt. The French navy was also successful at interdicting and seizing arms shipments at sea. For example, on January 18th, 1958, the French seized a Yugoslav ship destined for Morocco carrying 55 tons of arms and 95 tons of ammunition, which was enough to equip up to six ALN battalions.[29] Nonetheless, by year's end, the ALN was boasting a total force of nearly 130,000 of which 40,000 were uniformed fighters, and about 50% having undergone formal training in Tunisian bases.[30] At one point at least a thousand weapons were pouring across the border per month greatly enhancing the ALN's ability to regenerate combat strength and operations.[31] Although Morocco

[27] Ibid., 90..
[28] Ibid., 85.
[29] Ibid., 98.
[30] O'Ballance. 87.
[31] Horne. 230.

provided similar haven, Tunisia became the primary center of FLN activities providing:[32]

1. The most direct and secure route for arms and supplies
2. Convenient sanctuary for training and preparation of operations
3. A military and political headquarters in exile
4. A legitimate state supporter at international forums
5. A center for political negotiation with the French

Had it not been for this sanctuary, it is likely that the FLN would have collapsed and been decisively defeated before the fall of the Fourth Republic.[33] This explosion in armed fighters dictated drastic actions by the French beyond mere diplomatic maneuvering or France risked being overwhelmed.

Seeing the brewing cross-border crisis throughout 1957, the French undertook construction of a massive physical barrier between Tunisia and Algeria. The barrier was completed in September of 1957 and was a 200 mile long obstacle of mines, fencing, and electrified wire that became known as the Morice Barrage after the then French Minister of Defense.[34] At the time this border was state of the art and provided the French warning of any breaches to which a mobile quick react force could be dispatched.

This barrier was a top priority and at its height was actively patrolled and defended by as many as 80,000 personnel consisting of both French and Algerian troops, but later that number was significantly reduced.

---

[32] Ibid., 248.
[33] Ibid., 248.
[34] O'Ballance. 92.

The line was also pre-sighted for artillery fire and leveraged sensors and ground radar to alert the French to approaching enemy. At the heart of the defensive barrier was an eight-foot high electric fence through which ran a lethal 5,000 volts of current backed up by a trail that allowed highly mobile armored, dismounted, and motorized forces to easily patrol and quickly react to any infiltration of the line.[35] Once established, this border was extremely effective at cutting off ALN movement.

It was so effective, that the ALN resorted to desperation tactics of mass wave attacks to breach the border in order to get fighters into Algeria.[36] This allowed the French to devastate large numbers of insurgents in engagement after engagement. For example, on November 4th, 45 insurgents were killed near Tebessa, on December 1st, 48 insurgents were killed trying to cross near Souk-Ahras, on December 7th, 121 were killed near El-Ma-El-Aboid, on December 10th, 75 were killed near Guelma, and on December 19th, more than 70 were killed again near Tebessa.[37] This border was so effective, that the French Defense Minister was comfortable enough to announce the drawdown of 70,000 French forces in Algeria by the end of 1958 and cut the minimum military service time by three months.[38]

The beginning of 1958 brought a marked increase in the size of ALN forces trying to cross into Algeria from Tunisia. The ALN desperately needed to infiltrate fighters for a planned spring offensive and as such took

---

35 Horne. 264.
36 O'Ballance. 92.
37 Ibid., 92.
38 O'Ballance. 93.

great risks and suffered heavily along the Morice Line. More than 17,000 rifles, 380 machine guns, 296 automatic rifles, 190 bazookas, 30 mortars, and over 100 million rounds of ammunition were trapped in Tunisia preventing them from being distributed to the Wilayas that were becoming critically low on supplies.[39] Some of these operations included ambushing French patrols from the Tunisian border and eventually shooting down a plane.

This culminated on February 8th when a second French plane was hit by fire from the Tunisian village of Sakiet. In response, the French launched a cross border air raid that bombed the village of Sakiet, which utterly destroyed it.[40] This incident proved to backfire on the French and internationalize the incident in favor of the FLN. After seeing the less than desirable results, the French for the rest of the conflict, swore off what amounted to ineffective cross-border raids and stuck to a strong and much more effective defense of the border from the Algerian side. In response to the retrenched French defenses along the border, the insurgents tried a host of tactics, techniques, and procedures (TTPs) for breaching the obstacle belt, but met with continued failure. For example, attempts at cutting the electrified fence immediately alerted the French to the location of the breach and brought artillery bombardment followed by French reactionary forces flown in by helicopter and supported by ground units.[41] Attempts to breach the line with explosives also failed as the line was too wide and did not overcome the "no man's land" that extended

[39] Horne. 264.
[40] Ibid., 249.
[41] O'Ballance. 117.

sometimes up to 30 miles along the border where anything that moved was shot on sight.[42]

Attempts were also made to go around the line and infiltrate through the Great Erg sand deserts of the Sahara. These too failed as they proved easy for the French to locate and intercept in the wide open empty desert quarters.[43] The ALN continued in vain to break through the line throughout the winter and spring with increasing numbers of forces. This resulted in a staggering number of casualties that climaxed during the largest battle of the war between April 28th and May 3rd near Souk-Ahras where more than 1,000 ALN fighters tried to overwhelm the defenses. Instead, they were crushed sustaining casualties of more than 500 killed and a hundred captured.[44]

Even though a large number of these fighters did manage to breach the line and cross into Algeria, the rapid helicopter mobile French response quickly ran down, surrounded, and destroyed the insurgents before they could escape and link up with ALN units inside Algeria.[45] By the end of June, what came to be known as the "Battle of the Frontier" was overwhelmingly decided in the favor of the French. The ALN, having suffered 23,534 killed or captured since January attempting to breach the line, was crippled.[46] Further, more than 10,000 trained ALN fighters were "trapped" in Tunisia and unable to reenter Algeria to fight.[47] This effective cordon maintained by up to 30,000 French was maintained to the very end of the war. As a result, the

---

[42] Ibid., 118.
[43] Horne. 265.
[44] O'Ballance. 119.
[45] Horne. 266.
[46] O'Ballance. 120.
[47] Ibid., 120.

ALN was forced to return to much more limited guerilla style attacks providing the French the opportunity to consolidate their gains internally in Algeria. The ALN also had to shift their operations to Morocco and resort to smuggling arms and fighters into Algeria by a much longer and very dangerous southern route.[48]

The Morice Line was not without its critics. Colonel Roger Trinquier was one such outspoken person that argued the fence suffered from the same disadvantages of all fixed fortifications and a defensive mindset.[49] For example, the FLN was still openly operating in full sight of French forces across the Tunisian border with impunity. They also launched attacks from within Tunisia against the fixed positions taking a small, but steady toll that lowered French morale. Colonel Trinquier argued instead, the only way to decisively deal with the FLN was to strike in full force into foreign territory as was the norm he argued, of the pacification wars of the 19th century.[50] Nonetheless, by the end of 1958, the French were able to acknowledge they were turning the tide of the insurgency and winning for the first time on all fronts militarily.

General Maurice Challe, appointed by General de Gaulle, took over French operations in Algeria in 1959. One of his first priorities was to enhance the highly effective border security by increasing it along the Moroccan border as well as systematically eliminating the remaining resistance inside Algeria with highly mobile commando teams.[51] By spring, these operations had swept large areas in the mountains of remaining

---

[48] Ibid., 129.
[49] Horne. 266.
[50] Ibid., 266-267.
[51] O'Ballance. 131-132.

pockets of insurgents, which were becoming decisive since the ALN could not infiltrate replacements across the border in large numbers.[52]

In fact, the ALN in desperation had to begin forcing fighters to attempt to cross the border that by this time, now contained a second line along the Tunisian border.[53] Throughout the summer and fall, General Challe pressed his offensive and continued to capture and kill thousands of insurgents "trapped" in Algeria at a rate of about 500 fighters per day by very effectively cordoning an area and then clearing it.[54]

This had a telling effect on curtailing insurgent operations, but the ALN still managed to carry out terrorist style attacks to "show" it was still a force. Concurrently, what today may be termed "hearts and minds" initiatives were also conducted and enlarged that focused on improving the economic conditions for Muslim Algerians and conciliatory measures to include offers of amnesty and clemency.[55] However, these initiatives at this point seemed to have little, if any effect on the FLN, but did anger the Army.[56] France during this period also secured gains in its "political border control" initiatives. Unlike the physical acts of building and patrolling a barrier, the political acts consisted of diplomatic protests and blacklisting countries that supplied arms to the FLN. Specific threats to expose Italian and West German firms as well as Swiss bank financing precipitated a near total stoppage of arms deliveries from European as well as Arab countries.[57]

---

[52] Ibid., 133.
[53] Ibid., 133.
[54] Ibid., 134-135.
[55] Horne. 340-341..
[56] Ibid., 341.
[57] O'Ballance. 139.

Generally speaking, from late 1959 on, the ALN generally was only capable of small scale, guerilla style attacks. Even as late as 1961, ALN units were still trying to break through the Moroccan frontier, which by this point was defended in a manner similar to the Tunisian border making it futile. This led to entire ALN units being wiped out, and an official claim by France that the ALN was not able to penetrate more than 12 miles beyond the Moroccan border before they were detected and wiped out.[58]

By this point in the insurgency, the long term effects of being cut off, even in a country as large as Algeria were having a telling effect on the ALN. In particular, the acute shortage of weapons and ammunition made all but the most minimal resistance futile, and led to extremely low morale amongst the ALN. This also directly contributed to a growing inability to recruit new fighters, which meant the French had achieved decisive effects upon the ALN. This was perhaps all moot though after de Gaulle's September 16th, 1959 speech, in which he spoke of Algerian "self-determination."[59] Once de Gaulle presented this option, the military events ceased to matter and political events overtook the situation, pushing it irreversibly toward a final cessation of hostilities. As such, the ALN was effectively neutralized and shifted focus to political ends, which in March of 1962, culminated in the signing of the Evian accords that ended the conflict.

---

[58] Ibid., 164.

[59] Horne. 346.

Analysis

The conflict that began with few weapons and less than overwhelming support eventually grew to claim over a hundred thousand lives (nearer to a million by some accounts) and ended the French Fourth Republic. The collateral damage the civilian population suffered was enormous. Further, the political mishandling of the war nearly led to a complete collapse of the French government and long term damage to France's prestige.

Although the conflict was ultimately politically decided in favor of the insurgents, few would argue that the French did not achieve its military goals of suppressing the rebellion. To this point, the Morice Line and the French ability to lock down Algeria's borders, was a key critical element to the military success. Until the border was secured, the French were fighting a losing battle as the insurgents grew in numbers, training, and arms. From the moment the border was sealed, the insurgency began to suffer quantitatively and qualitatively.

For the first time, the French could achieve decisive results in clearance operations knowing that the insurgents were no longer able to effectively re-arm and supply. The French however, through brutal tactics that included widespread use of torture, indiscriminate killing of civilians, and forced relocation of others to what amounted to squalid refugee camps, lost the moral high ground and support of a population that could have been turned against the FLN. The French population also became tired of the conflict and appalled by the revelations of the military activities, which ultimately convinced de Gaulle of the futility of the affair. This opened the door to "self-determination" and the ultimate culmination of a political agreement granting Algeria independence. As such, the Algerian case strongly conforms to the hypothesis. Specifically, it was clear that effective border security was a necessary condition for neutralizing the FLN, but not sufficient on its own to win the war.

From this case study, many lessons can be drawn for COIN strategy as it relates to border security. Most important to the thesis argument is that until the flow of arms and personnel were cut off, the insurgents were not only able to sustain operations, but were able to greatly expand them in spite of large-scale French operations. Until the French sealed the border effectively trapping the remaining insurgents inside Algeria and preventing the insurgents outside of Algeria to help, the operations were futile. From that point on, the insurgency inside Algeria was crippled and effectively withered without its cross-border support.

Operationally, this was seen when the French employed countless cordon-and-search operations that were initially ineffective. It wasn't until the French began conducting massive operations to ensure the cordon was effectively established before closing in on and clearing the insurgent strongholds that they were able to prevent insurgent escape. This coincided with the border being closed so these gains could be driven home for decisive results because the insurgents were unable to resupply. Second, the border obstacles no doubt involved large numbers of local labor and showed the French commitment to stay the course. David Kilcullen identifies this consistency and long term commitment as a critical counterinsurgent tool in winning over the population.[60]

---

[60] Kilcullen. 93-95.

Further, as Kilcullen identifies in his Afghanistan case study on a road building project in Kunar Province of Afghanistan, these type projects become an irresistible target for the insurgent and challenge local insurgent legitimacy. Kilcullen points out that this in fact forces the insurgent to come to the counterinsurgent where the counterinsurgent can then bring to bear his superior firepower and destroy the insurgent on his terms vice endlessly chasing the insurgent in vain on his terms.[61]

This was independently confirmed by a Marine Corps officer operating in Regional Command East (Afghanistan). The Marine officer, now an instructor at the Marine Corps Expeditionary Warfare School, testified to the fact that Border Check Points (BCPs) were irresistible targets that in his opinion threatened insurgent freedom of movement and provided them propaganda opportunities regardless of the tactical outcome. He went on to state that this wasn't a bad thing because it was one of the few times his forces were able to "fix and destroy" the insurgents.

This appeared to be the case in Algeria not just for the reasons Kilcullen notes, but also because it presented a very real operational hurdle to the insurgent that had to be overcome for survival. This forced the insurgent into a conventional confrontation with the far superior French forces, which had a devastating impact on insurgent morale and numbers. Third, to make the border obstacles more effective, the French developed "no-man's" land areas along the frontier so that any infiltrators could be easily interdicted.

To accomplish this, the French made wide scale use of population resettlement, which allowed them to isolate and remove and destroy insurgent strongholds in

---

[61] Ibid., 95-97.

difficult regions. This has been an effective COIN tactic employed in many struggles such as Vietnam to good effect. However, the failure of the French to ensure good conditions were maintained in these camps ultimately succeeded in alienating the population and quite likely turning more people to the side of the FLN than the FLN could have done on its own.

Fourth, limited cross-border raids not only proved to be ineffective, but caused much more damage to the counterinsurgent at the strategic level than any tactical gain could justify. Perhaps Roger Trinquier was correct in that to completely destroy the FLN, one would have to conduct full scale invasion of neighboring sanctuary, but now just as it was then, this idea was neither militarily nor politically possible. As such, the French learned all too late that it would have been better never to launch a cross-border attack if they were to be only of a limited nature due to the strategic political blowback.

The Algerian case study is also valuable in dispelling many critiques put forth about the effectiveness of border security. The first comes from Colonel Trinquier, who pointed out that the Morice Barrage suffered from problems of static defenses. He was correct in noting this in so far as the FLN did try to just "go around" it, but failed to acknowledge that this too proved to not be a viable option. He also suggested that troops deployed in static locations would be complacent and morale would be low. This is an interesting perspective in that the troops deployed to the border were some of the most elite and were continually on the move hunting down, surrounding, and destroying large bodies of FLN fighters attempting infiltration.

They were anything, but "static" and any suggestion of a defensive posture looked more like an offense in reality. In fact they fought the largest battles of the war and inflicted arguably the heaviest casualties on the FLN. Extremely important to also note is that most of these fighters were killed or captured cleanly without causing civilian collateral damage that plagued operations inside Algeria. The morale of troops did suffer from repeated harassing attacks, but this was more likely a problem of poor leadership and troop rotation and can in no way be compared to the incredible effect it had not only mentally on the morale of the insurgents, but more tangibly in their ability to wage war.

Second, skeptics of border security will repeatedly note that it would be too difficult and too expensive to implement and that even if it was, the expenditure wouldn't be justified because it could never be effective. Not only were the French able to rapidly build a barrier through hundreds of miles of both mountains and desert, but proved that this was a highly effective COIN tool. It should be noted that often in the open desert, the French did not even need to build a barrier and instead relied on radar early warning systems and observation coupled with a helicopter transported react force fully capable of chasing down any infiltrators attempting to move undetected through the vast open desert regions.

Although the border barrier did require a large investment of capital to construct, the necessity to build it and the critical role it played more than justified the cost. In fact, based on the evidence, it could be convincingly argued that if the French were unwilling to dedicate the time and resources to securing the border, it would have been near impossible to defeat the insurgency. Without stopping the flow of insurgents and weapons, it would have simply been futile to contest the FLN.

The Algeria case study confirms much of what was observed from analysis of the Greek study in regards to implications for Afghanistan. First, it is clear a militarily far superior force can be checked and defeated by an insurgent force deriving benefits from cross-border, state supported sanctuary. No matter how many times the internal areas of a country were cleared, the insurgents were able to quickly re-establish as long as they could withdraw to safety across the border during the operation and return at will at a new place of their choosing.

This robbed the counterinsurgent the chance to conduct decisive operations. Following this logic, it is unlikely that NATO clearance operations in the Helmand and Kandahar Provinces of Afghanistan will have any lasting effect. More importantly, these operations will be unable to achieve decisive results. Drawing directly from the Algerian case study, limited cross-border raids had a decidedly damaging effect at the strategic level for the French. This should be seen as a point of caution for the US, which has increased its use of targeted killings inside of Pakistan via unmanned drone strikes fomenting strong outrage and anti-American sentiments amongst Pakistanis.

In regards to the feasibility of implementation, the French were able to effectively secure a border of very similar length and topography in a very short period of time through both mountains and desert. This proves securing Afghanistan's border with Pakistan is feasible should NATO decide on that course of action. Although it would require a large re-tasking and deployment of the current forces in country, the effects would allow future operations to achieve more enduring results.

Relative to cost, the border could likely be secured with as little as $3.3 billion providing a real chance to win in Afghanistan. This rough estimate is derived simply by multiplying the maximum cost per mile of state-of-the-art border security and barriers employed by the US and Israel (about $2.2 million) times the number of miles along the Afghanistan-Pakistan border (~1,510 miles). Although large, in terms of the amount of money the US spends daily in Iraq and Afghanistan, this is pennies. For example, the US spent more money in two weeks in Iraq at its height, than it would cost to seal the border. Even the $3 billion spent for the "cash for clunkers" program in 2009 could have paid the price.

The US, by controlling the borders, would also find it reaps second order effects that also damage the insurgents' ability to operate, while bolstering the legitimacy of the state. In particular, securing the borders would cut down, and in some cases, eliminate illicit cross border smuggling, which would drastically affect the ability to move drugs. This dent placed in the narcotics trafficking would cut the amount of money flowing to the Taliban.

Chapter Three
Tibet and Israel

Numerous examples of effective border security, in past and present conflict zones such as North Ireland and Kashmir abound, but a couple select examples are worth more detailed mention. First, is the case of Tibet that provides perspectives on the effectiveness of border security from the insurgent. Second, is the case of Israel, which probably more so than any other state has made extremely effective use of border security to fight terrorism emanating from an insurgent Palestinian population.

Tibet

The war for Tibet's autonomy from China and the CIA's role in insurgency is only now coming to light due to declassified files. Although the data is still limited, the Tibet example provides a unique perspective of effective border security from the insurgent perspective. During this struggle, the Tibetans received substantial state support from the CIA and cross-border sanctuary in India and Nepal. However, we know from CIA case officer memoires, declassified US files, and Tibetan insurgents that the Chinese relentlessly increased the pressure on the Tibetan insurgency, which eventually led to a successful Chinese political-military effort to secure the border and remove the insurgent sanctuary. Physically, the Chinese effectively sealed off the mountain passes connecting Tibet with the insurgent sanctuaries.

This began in 1959 with the deployment of 100,000 Chinese soldiers from the People's Liberation Army

(PLA) and took two years to complete.[1] At that point, the insurgents were no longer able to infiltrate and supply the insurgency inside China. In response, the Tibetan insurgents moved their main base of operations to a camp known as "Mustang," which was a high mountain, cross-border stronghold, 150 miles northwest of Katmandu, Nepal.[2] However, by 1960, the PLA had secured its control of the border by protecting the high mountain passes so effectively, it was no longer possible for the insurgents to maintain effective resistance beyond limited and inconsequential raids.[3]

This meant the resistance was neutralized, but it did not mean the complete end of the organization, until the political events finally removed the sanctuary. Specifically, the Chinese warming with the US politically compromised the CIA's on-going assistance to the rebels. This meant the US finally canceled its assistance to the Tibetan insurgents still residing inside Nepal. The final blow came after the Chinese government placed such heavy pressure on the king of Nepal, that in 1974, King Birendra Bikram ordered the patrolling of its side of the border and 10,000 Nepalese soldiers to march against Mustang.[4] This military force overwhelmed the camp and the remaining insurgents were captured or killed. This culmination of political and military events to secure the Chinese borders ultimately allowed the Chinese to decisively destroy the insurgency.

---

[1] Prados, John. *Presidents' Secret Wars: CIA and Pentagon Covert Operations since World War II.* New York: W. Morrow, 1986. 164-65. Print.

[2] Prados. 168.

[3] Prados. 169.

[4] Prados. 169-170.

Israel

Israel is a state that is continually attacked by both internal and external sources, which includes insurgent factions such as Hezbollah and Hamas that leverage terrorism as a means of resistance. As a result, internal and external border security has become central to Israeli security strategy. Israel's effective use of a sophisticated border security system to deal with terrorism as well as insurgency suggests that border security has much wider implications beyond pure COIN operations and makes it of key interest for study.

The official purpose of the Israeli Security Fence, according to the Israeli Ministry of Defense, is "to provide a response to the threats posed to the State of Israel and protect its population from the threat of terror and criminal activity."[5] The fence is designed to serve as a multilayered defense that includes:[6]

1. A ditch and a pyramid shaped stack of six coils of barbed wire on one side and barbed wire on the opposite side.
2. A path enabling the patrol of Israeli Defense Forces (IDF) on both sides.
3. An intrusion-detection fence, with sensors to warn of any incursion.
4. A smoothed strip of sand that runs parallel to the fence, to detect footprints.

---

5 "Isreal's Security Fence." *Israeli Ministry of Defense*. 2007. Web. <http://www.securityfence.mod.gov.il/Pages/ENG/questions.htm>.

6 "Isreal's Security Fence." *Israeli Ministry of Defense*. 2007. Web. <http://www.securityfence.mod.gov.il/Pages/ENG/questions.htm>.

5. A solid barrier system in select areas (less than 6% of the total fence) to prevent sniper fire along urban areas and roads.
6. Observation towers, checkpoints, and crossing points that employ a host of technologies to screen for weapons and identify individuals, including collection and verification of biometric data.

These characteristics are illustrated in figures 1 and 2 below.

Figure 1.[7]

[7] "Isreal's Security Fence." *Israeli Ministry of Defense*. 2007. Web. <http://www.securityfence.mod.gov.il/Pages/ENG/questions.htm>.

Figure 2.[8]

In the early 1990s, Palestinian insurgent activity became more violent and included a growing incidence of terrorist attacks using suicide bombers inside Israel. Until this time, Israeli counterterrorism efforts relied mainly on intelligence and policing operations including targeted killings. However, these methods proved ineffective against the rise in violent attacks.

This prompted the Israeli government to take decisive actions to check the unhindered movement of terrorists, weapons, and money by containing Palestinian population areas. Led by then-Prime Minister Yitzhak Rabin, the Israelis, in 1992, modified their counterterrorism strategy to include internal border

---

[8] "Israeli Security Fence." *Israeli Ministry of Defense*. 2007. Web. 16 Dec. 2010. <http://www.google.com/imgres?imgurl=http://www.securityfence.mod.gov.il/pages/ENG/images/operational1.jpg&imgrefurl=http://www.securityfence.mod.gov.il/pages/ENG/operational.htm&usg=__uSNAFWv_nuvkN9kspZ-gAVCaQoQ=&h=394&w=680&sz=60&hl=en&start=17&zoom=1&tbnid=P2x_ww6rxtKJ1M:&tbnh=81&tbnw=139&prev=/images?q=israeli+security+fence&um=1&hl=en&sa=N&tbs=isch:1&um=1&itbs=1>..

security.[9] The Israeli model was designed to stop terrorism by denying the interior of Israel to terrorists while also cutting outside support to the insurgent factions within Israel's borders. Academics such as Robert Trager and Dessislava Zagorcheva supported this strategy and argued that through border security and hardening targets a country can effectively deter terrorists by reducing the effectiveness of their tactics.[10]

The new strategy first materialized as the Israeli Security Fence, which was begun in 1994 and eventually cordoned Gaza off from the rest of Israel, an area approximately 40 kilometers long. The isolation of Gaza followed what has become a predictable pattern in the employment of border security: violence grows unchecked, then effective border security is implemented, the insurgents launch ineffective resistance, and finally the insurgency is either decisively defeated or forced to ask for a ceasefire.

In the case of Gaza, the Palestinians agreed to negotiations at Camp David in 2000. Since the fence was rebuilt in 2001 after it was partially destroyed during the First Intifada, there has been an unprecedented prevention of terrorist attacks with virtually no incidence of intrusions beyond the border.[11] As a result, "hundreds of attempted infiltrations were thwarted inside the buffer zone before the terrorists ever reached

---

9 Makovsky, David. "How to Build a Security Fence." *Foreign Affairs* 83 (2004): 50-64. Print.

10 Trager, Robert, and Dessislava P. Zagorcheva. "Deterring Terrorism: It Can Be Done." *International Security* Winter 30.3 (2005-2006): 87-123. *Project MUSE.* Web. <http://muse.jhu.edu/login?uri=/journals/international_security/v030/30.3trager.html>.

11 Almog, Doron. "The West Bank Fence: A Vital Component in Israel's Strategy of Defense." *The Washington Institute for Near East Policy* (2004): 8. Print.

the electronic fence."[12] In fact, the only noteworthy attack successfully launched from Gaza after the border fence was constructed occurred on June 25, 2006. In this incident, Palestinian insurgents dug a tunnel under the fence and breached the border. Two Israeli soldiers were killed and one captured in the ensuing attack, which notably was contained at the border and did not threaten the civilian population.[13] Since then, Israel has enhanced its tunnel detection capabilities and this has not been repeated.

Aside from this incident and generally inaccurate rocket attacks, the rebuilt fence has eliminated successful suicide attacks emanating from Gaza, achieving an unprecedented 100% containment of terrorists by some accounts.[14] Since the tunnel incident in 2006, all other attempts to circumvent the border's effectiveness have been unsuccessful and insurgents inside Gaza seeking to carry out terrorist attacks against Israel have utterly failed. Figure 3 illustrates the reduction in Israeli fatalities in relation to the construction of the security fence around Gaza.

---

[12] Kershner, Isabel. *Barrier: the Seam of the Israeli-Palestinian Conflict.* New York: Palgrave Macmillan, 2005. 161. Print.

[13] Vitka, William. "Palestinian Militants Attack Border - CBS News." *Breaking News Headlines: Business, Entertainment & World News - CBS News.* 25 June 2006. Web. 16 Dec. 2010. <http://www.cbsnews.com/stories/2006/06/25/world/main1749360.shtml>.

[14] Almog. 8.

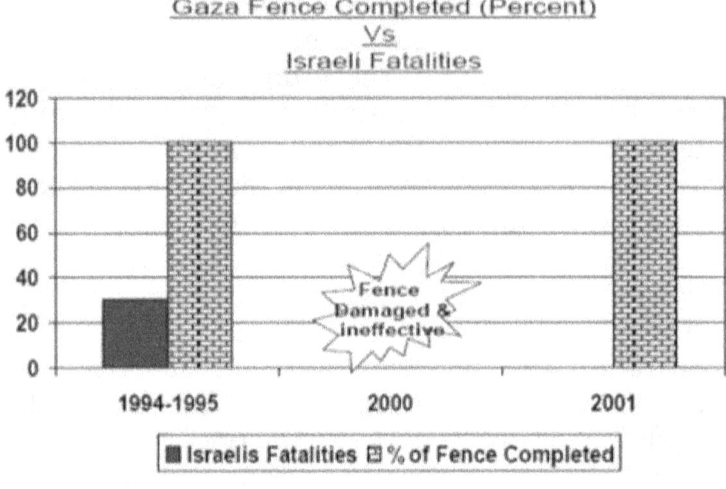

Figure 3.[15]

The next wave of violence to hit Israel coincided with the beginning of the Second Intifada, in September of 2000. On February 21, 2002, following a rash of suicide bombings launched primarily from the West Bank, Prime Minister Ariel Sharon declared his support for a new barrier around the West Bank.[16] Although contentious, the Israeli government supported the plan after Palestinian terrorists killed 80 Israelis and wounded 600 others in twelve different suicide attacks.[17] On April 14, 2002, Sharon's cabinet approved construction of three "buffer zones," which began within

[15] Kershner, Isabel. "Israel Security Fence." *Israeli Ministry of Defense*. 2006. Web. <http://www.securityfence.mod.gov.il/pages/eng/news.htm>.

[16] Thein, Ben. "Is Israel's Security Barrier Unique." *The Middle East Quarterly* Fall (2004). Web. <http://www.meforum.org/652/is-israels-security-barrier-unique#_ftn5#_ftn5>.

[17] Thein.

two months of approval.[18] Although skeptics point out other methods of countering the growing insurgent terrorist activity were what stemmed the violence, the Israeli Supreme Court's ruling on the legality of the barriers, best states the need for border security, and the failure of all other methods to stop the terrorist attacks. In the ruling, the court states:

> These terrorist acts committed by the Palestinian side have led Israel to take security steps of various levels of severity. Thus, the government, for example, decided upon various military operations, such as operation "Defensive Wall" (March 2002) and operation "Determined Path" (June 2002). The objective of these military actions was to defeat the Palestinian terrorist infrastructure and to prevent reoccurrence of terror attacks . . . These combat operations – which are not regular police operations, rather bear all the characteristics of armed conflict – did not provide a sufficient answer to the immediate need to stop the severe acts of terrorism. The Committee of Ministers on National Security considered a series of steps intended to prevent additional acts of terrorism and to deter potential terrorists from committing such acts . . . Despite all these measures, the terror did not come to an end. The attacks did not cease. Innocent people paid with both life and limb. This is the background behind the decision to construct the separation fence (Id., at p. 815).[19]

---

[18] Thein.

[19] Israel High Court Ruling Docket H.C.J. 7957/04: International Legality of the Security Fence and Sections near Alfei Menashe. Supreme Court of Israel. 15 Sept. 2005. Web. <http://www.zionism-israel.com/hdoc/High_Court_Fence.htm>.

The effects of the border fence around the West Bank, like the Gaza fence, demonstrated a dramatic decrease in attacks. Specifically, in the 34 months since the beginning of the Second Intifada prior to the erection of the fence, terrorists succeeded in carrying out 73 attacks, in which 293 Israelis were killed and 1,950 were wounded. [20] In comparison, from August 2003 to June 30, 2004, terrorists in the Northern West Bank managed to carry out only three suicide attacks in Israel (all of which occurred in the second half of 2003), killing 26 Israelis and wounding 76.[21] On two occasions, the terrorists infiltrated through areas in the Northern West Bank where the fence was still under construction. On the third occasion a female suicide bomber infiltrated into Israel through the Bartah crossing by exploiting leniency towards women and using a Jordanian passport. [22] These incidents demonstrate a significant decrease in attacks that chronologically corresponds directly with the construction of the barrier.

The correlation between improved security and border construction is also evidenced by the decline in the number of suicide bombing attacks relative to the increase in miles of secured border around the West Bank. During 2004, fourteen suicide bombing attacks were carried out, compared with seven in 2005 and four in 2006. Of the attacks that occurred in 2006, three were carried out by the Palestinian Islamic Jihad (PIJ) and one by Fatah. Hamas did not conduct any attacks in 2006. [23] By this time, terrorists were simply unable to

---

[20] "IDF Spokesperson." Interview. 7 July 2004. Web. <http://www.imra.org.il/story.php3?id=21411>.

[21] IDF Spokesperson.

[22] IDF Spokesperson.

[23] *Anti-Israeli Terrorism, 2006: Data, Analysis and Trends*. Rep. Intelligence and Terrorism Information Center at the Israeli

cross the security border. Most of the suicide bombers infiltrated, or intended to infiltrate, through the area around Jerusalem, exploiting the weak spot in the security fence. [24] The charts below illustrate this point.[25]

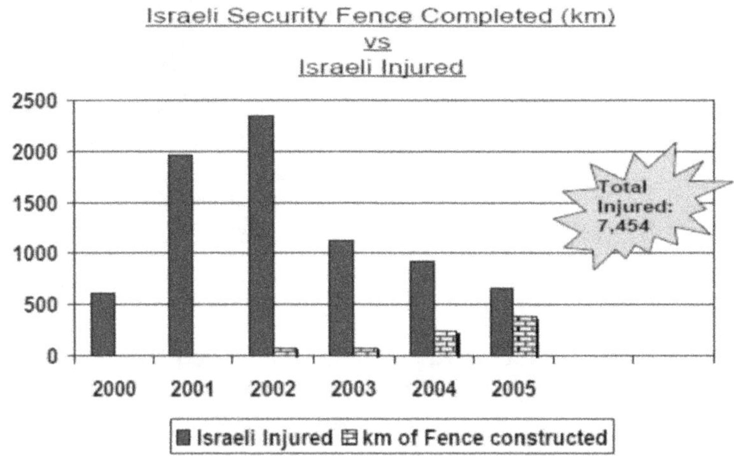

Heritage & Commemoration Center, 2007. Web. <http://www.terrorism-info.org.il/malam_multimedia/English/eng_n/pdf/terrorism_2006e.pdf.>.

24 *Anti-Israeli Terrorism, 2006: Data, Analysis and Trends.* Rep. Intelligence and Terrorism Information Center at the Israeli Heritage & Commemoration Center, 2007. Web. <http://www.terrorism-info.org.il/malam_multimedia/English/eng_n/pdf/terrorism_2006e.pdf.>.

25 *Anti-Israeli Terrorism, 2006: Data, Analysis and Trends.* Rep. Intelligence and Terrorism Information Center at the Israeli Heritage & Commemoration Center, 2007. Web. <http://www.terrorism-info.org.il/malam_multimedia/English/eng_n/pdf/terrorism_2006e.pdf.>.

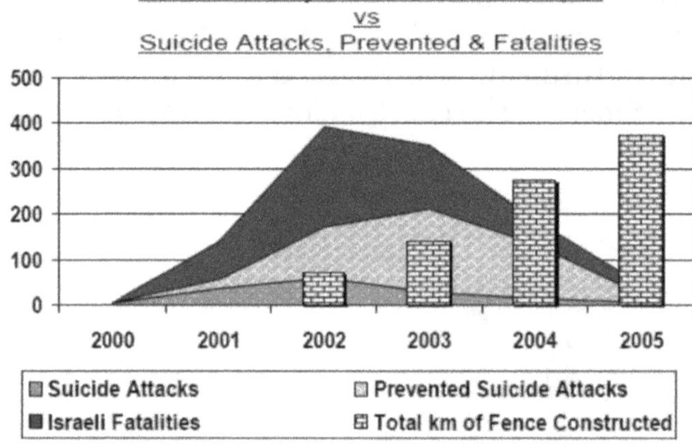

Figure 4.[26]

Securing Israel from attack also requires Israel maintain effective coastal border security to prevent external state support from reaching insurgent organizations such as Hamas, PIJ, and the Popular Front for the Liberation of Palestine (PFLP). This is particularly important since Gaza lies along the coast of the Mediterranean Sea. The coastal border has been a point of access for aid to Palestinian terrorist groups from states such as Iran, who provide training, weapons, and financial support.[27] To prevent this aid from crossing the coastal border, the Israeli Navy and Coast Guard have interdicted large amounts of weapons at sea, effectively cutting the last major mode of resupply Palestinian groups had available after the security fences

[26] Erlich, Reuven. "Suicide Bombing Terrorism During the Current Israeli-Palestinian Confrontation September 2000-December 2005." *Intelligence & Terrorism Information Center :: Homepage.* 2006. Web. 16 Dec. 2010. http://www.terrorism-info.org.il/site/home/default.asp; Israeli Ministry of Defense.

[27] *Erased in a Moment: Suicide Bombings Against Israeli Civilians.* Rep. USA: Human Rights Watch, 2002. 94-107. Print.

cut land routes. Recognizing the importance of coastal security, Israel made interdiction of arms entering Gaza by sea, one of its criteria for a ceasefire, during the 2009 offensive in Gaza (Operation Cast Lead).[28] Based on the large reduction in rocket attacks from Gaza since the 2009 offensive, one can argue this cordon by sea has indeed succeeded in cutting the last major resupply route to the Palestinian insurgents. Thus, this naval aspect of effective border security has shown itself to be critical in Algeria, Greece, and Israel.

Evidence further cementing the case that it was indeed the effectiveness of the internal border security that was primarily responsible for the reduction in terrorist violence is provided by the insurgents. PIJ leader, Ramadan Abdallah Shalah, when interviewed by the Qatari newspaper, Al-Sharq, stated during the Second Intifada rocket fire had replaced the previous stage of suicide bombing attacks because the enemy [i.e., Israel] had found ways and means to protect itself from such attacks:[29]

> ... For example, they built a separation fence in the West Bank. We do not deny that it limits the ability of the resistance [i.e., the terrorist organizations] to arrive deep within [Israeli territory] to carry out suicide bombing attacks, but the resistance has not

---

[28] Radia, Kirit. "New Details on U.S.-Israel Pact to Block Arms to Hamas - Political Radar." *Political Radar*. ABC News, 15 Jan. 2009. Web. 16 Dec. 2010. <http://blogs.abcnews.com/politicalradar/2009/01/new-details-on.html>.

[29] *Anti-Israeli Terrorism, 2006: Data, Analysis and Trends*. Rep. Intelligence and Terrorism Information Center at the Israeli Heritage & Commemoration Center, 2007. Web. <http://www.terrorism-info.org.il/malam_multimedia/English/eng_n/pdf/terrorism_2006e.pdf .>. *Al-Sharq, March 23, 2008.*

> surrendered or become helpless, and is looking for other ways to cope with the requirements of every stage [of the intifada]...

Abdallah Shalah went further and told Hezbollah's Al-Manar TV that the terrorist organizations had no intention of abandoning suicide bombing attacks, but that their timing and the possibility of carrying them out from the West Bank depended on other factors. “For example,” he said, “there is the separation fence which is an obstacle to the resistance [i.e., the terrorist organizations], and if it were not there, the situation would be entirely different.”[30]

Mousa Abu Marzouq, deputy chairman of Hamas' political bureau, also made a similar statement on why suicide bombing activity had decreased following the election of the Hamas government. He said “[carrying out] such attacks are made difficult by the security fence and the gates surrounding West Bank residents.”[31]

Considering the correlation in time between the construction of internal border security measures and the reduction of violence twice over in Gaza and then the West Bank, statements by Israel’s Supreme Court and insurgent leaders, and the continued state of relative security in Israel, there is an indisputable connection

---

[30] *Anti-Israeli Terrorism, 2006: Data, Analysis and Trends*. Rep. Intelligence and Terrorism Information Center at the Israeli Heritage & Commemoration Center, 2007. Web. <http://www.terrorism-info.org.il/malam_multimedia/English/eng_n/pdf/terrorism_2006e.pdf .>. Al-Manar TV, November 11, 2006.

[31] *Anti-Israeli Terrorism, 2006: Data, Analysis and Trends*. Rep. Intelligence and Terrorism Information Center at the Israeli Heritage & Commemoration Center, 2007. Web. <http://www.terrorism-info.org.il/malam_multimedia/English/eng_n/pdf/terrorism_2006e.pdf .>. Abd al-Muaz Muhammad, Ikhwan Online, The Muslim Brotherhood Website, June 2, 2007.

between the physical barrier of the security fence and gaining an upper hand over insurgents. To this day, Israel has not had the rash of terrorist attacks that it suffered before border security was implemented around Gaza and the West Bank. Israel has continued construction of the security fence, and as of April 2006, the length of the barrier was approved by the Israeli government to be a total of 703 kilometers (436 miles) long. Regardless of political condemnation by human rights organizations, the security fence has received widespread Israeli public support for reducing incidents of suicide bombings to almost zero, and will likely remain a permanent fixture of Israel's security strategy.

Part 2

Border Security in Counterinsurgency

Chapter Four

Asymmetric Approaches to Border Security in COIN

Border security encompasses much more than just fences. As the Greek, Algerian, and Israeli case studies demonstrate, the counterinsurgent must use all resources at his disposal to ensure the borders are effectively secured. This combines both political and physical elements to maximize the effectiveness. However, when possible, the best situation is when the counterinsurgent can quickly maneuver to stop an insurgency by shutting their borders to deny sanctuary.

Asymmetric approaches have shown to provide some effectiveness in this regard. El Salvador provides the first example where neighboring Nicaragua would not cease its support for communist insurgents and the counterinsurgent devised a partially effective plan to mitigate the amount of cross-border support the insurgents received. The second, is the on-going and yet concluded conflict in Iraq, which provides a unique twist through "internal" border security.

El Salvador

Enough literature in English exists on the leftist insurgency in El Salvador to correlate the counter-insurgent victory with a border security plan. In this specific case, leftist rebels primarily derived support and sanctuary from cross-border bases supplied by the Soviets and Cubans in Nicaragua. The insurgency was largely unpopular in El Salvador, but lingered on, due in part to the inability to physically secure the border. As such, an asymmetric plan to secure the border was derived. This plan, headed up the by CIA, supported an insurgency inside Nicaragua from cross-border sanctuary primarily in Honduras.[1] From Honduras, the Contras were able to harass and destabilize the Leftist Sandinista regime in a significant enough measure to force Nicaragua to concentrate on securing itself domestically, which reduced the amount of support they were able to provide the insurgents operating in El Salvador.[2]

As a result, the El Salvadorian government gradually gained the upper hand until the collapse of the Soviet Union, which brought the end of state support and ultimately the end of the insurgency for the leftist rebels. A negotiated settlement was reached soon afterwards. This case is illustrative because it leverages an asymmetric approach to cutting insurgents off from the state supported sanctuary. Although not a clear cut case in easily quantifiable terms, the evidence does suggest that as the benefits derived from the Nicaraguan sanctuary diminished, so did the ability of the rebels to operate.

Iraq

Iraq seems to have presented a full host of border security options. In the cases of Saudi Arabia, Kuwait, Jordan, and Turkey, political interests aligned in a way that each respective country was independently

---

[1] Rabasa, Angel, Lesley A. Warner, Peter Chalk, Ivan Khilko, and Paraag Shukla. *Money in the Bank: Lessons Learned from past Counterinsurgency (COIN) Operations*. Santa Monica, CA: Rand, 2007. 48. Print.

[2] Rabasa, Angel, Lesley A. Warner, Peter Chalk, Ivan Khilko, and Paraag Shukla. 48.

motivated to seal their borders with Iraq. Although some infiltration (especially in the cases of Saudi Arabia and Jordan) occurred, these countries eventually constrained movement to such a low level, especially in the case of Jordan, that infiltration would not have a destabilizing threat to Iraq's sovereignty.

Further, Turkey and Saudi Arabia have also spent considerable resources on developing fortified borders to prevent insurgents from Iraq entering their countries. This left Syria and Iran presenting Iraq with a legitimate cross-border threat. However, both countries legitimately feared US military intervention and invasion if their support became too overt and destabilizing. This is different from a country like Pakistan, that has a reasonable expectation the US would not invade it, due in no small part to its nuclear arsenal (the US does provide billions of dollars in foreign aid to Pakistan that could be leveraged, but the US has not effectively wielded this bargaining chip). This allowed the US to pressure Iran and Syria in ways that it is unable to in a case such as Afghanistan.

Further, the US military and Iraqis conducted border security and interdiction operations to a noticeable effect, even though they did not have a large amount of forces dedicated to this task. This was possible due to the increased ease at which an infiltrator could be identified and apprehended in Iraq's vast unpopulated desert regions along its borders.

The US inadvertently undertook what this thesis refers to as an "internal border security" initiative with its "Inkblot Strategy" to control population centers. This had the effect of internally sealing the borders of Iraq from insurgents, because it was extremely difficult to find suitable sanctuary in the open, under populated desert regions of Iraq outside the highly concentrated population centers. Under these conditions, it is more efficient to employ troops in an internal security role vice trying to guard long open areas of desert where large numbers of forces would be tied down and never do anything.

This ultimately played a decidedly unappreciated and unrealized role in the combined US-Iraq effort to gain the upper hand on the insurgency and suppress it to an acceptable level. Evidence of this appears in FM 3-24's discussion of how Tal Afar was secured. FM 3-24 specifically describes how a berm was built around the city to cut it off from insurgents attempting to infiltrate while the city itself was systematically cleared and secured by US and Iraq forces. Through the lens of internal border security this action appears to have been a critical element; however, the benefits of this went unrealized by the US military and were not institutionalized in its doctrine or overall strategy.

The second example comes from how Sadr City was secured and provides detailed powerful evidence of the impact of internal border security. The battle for Sadr City began in April 2004, after the Coalition Provisional Authority moved to constrain the power of Muqtada Al-Sadr and his Sadrist Movement.[3] The outbreak of violence led to continued conflict between insurgents and US forces on a sustained basis that included occasional shelling of the Green Zone and heavy use of improvised explosive devices (IEDs) along roadways. By 2008, the fighting had lasted four years, and the US

---

3 "CBC News - World - Fighting Kills Dozens of Iraqis, 8 U.S. Soldiers." *CBC.ca - Canadian News Sports Entertainment Kids Docs Radio TV*. 6 Apr. 2004. Web. 16 Dec. 2010. <http://www.cbc.ca/world/story/2004/04/06/iraq_casualties040406.html>.

began erecting a wall inside of Sadr City on April 15th to stem the continuous flow of insurgents.[4]

This led to a major escalation in combat with insurgents launching repeated attacks over the next month with many insurgents being killed.[5] However, by mid-May, the US and Iraqi forces had gained the upper hand and by the end of the month, Iraqi forces had secured and occupied the city. Inside of roughly a month, fighting that had lasted four years, came to an abrupt end with a negotiated cease-fire.[6] Reports on exact casualties may never be known, but the US military estimated it killed 700 insurgents during the fighting at the cost of six Americans killed.[7] Again, just as in Tal Afar, the victory was not linked to the internal border security and the US military does not seem to have appreciated the value the barriers provided. Nonetheless, the evidence again suggests there is a strong correlation that these barriers were in fact a key, critical element to securing a victory within one month of implementation, which for four years had eluded coalition forces.

Adding to the evidence is the recurrent pattern of long term insurgent activity, implementation of effective border security, heavy fighting leading to unsustainable

---

[4] Gordon, Michael. "U.S. Begins Erecting Wall in Sadr City." *The New York Times*. 18 Apr. 2008. Web. <http://www.nytimes.com/2008/04/18/world/middleeast/18sadrcity.html?_r=4&ref=world&oref=slogin&oref=slogin>.

[5] "Siege of Sadr City." *Wikipedia, the Free Encyclopedia*. 14 Nov. 2010. Web. 16 Dec. 2010. <http://en.wikipedia.org/wiki/Siege_of_Sadr_City>.

[6] "Siege of Sadr City." *Wikipedia, the Free Encyclopedia*. 14 Nov. 2010. Web. 16 Dec. 2010. <http://en.wikipedia.org/wiki/Siege_of_Sadr_City>.

[7] Hodge, Nathan. "Gaza Ground Campaign Mirrors Battle of Sadr City? | Danger Room | Wired.com." *Wired.com, Danger Room*. 6 Jan. 2009. Web. 16 Dec. 2010. <http://www.wired.com/dangerroom/2009/01/gaza-ground-cam/>.

insurgent losses, and within a few months or less, a decisive result such as a negotiated settlement or collapse of insurgent resistance. This pattern has been repeated at least twice inside Iraq, was witnessed in Greece, was proven in Israel, and as will be discussed, was also demonstrated by the Soviets in Afghanistan.

The Soviet War in Afghanistan

To cross check the hypothesis, it is important to identify a strong case where there was a cross-border state supported insurgency and the counterinsurgent did not effectively implement border security. For the hypothesis to have merit, one would predict the counterinsurgent should generally lose in these situations, especially if the insurgent had strong state support. As such, a strong test was identified, where a superpower engaged in this type of conflict.

The case is the Soviet Union in Afghanistan (December 1979-February 1989). The Soviet superpower attempted to decisively defeat a poorly armed and equipped, ill trained, and generally disorganized band of insurgent groups, yet ultimately failed after almost a decade of conflict. This case provides an important context for analysis, because an initially very weak group of insurgents took on a world superpower, in the very same region the US/NATO is fighting today. Further, the war's duration, combatants (insurgent forces), sanctuary (Pakistan), and transnational nature of the insurgency are all so similar to the current US/NATO war in Afghanistan, that they approach what could be considered as controls during comparative analysis.

During the Soviet War in Afghanistan, an initially poorly trained, organized, and armed band of Islamic insurgent groups operating out of state supported sanctuary in Pakistan, defeated a superpower. These insurgents, known collectively as the Mujahedeen, eventually coalesced into seven major factions based in Pakistan.[8] Noteworthy is the long term survivability of these groups. In fact, many of the same factions and leaders, such as Gulbuddin Hekmatyar's Hizb-i-Islami Afghanistan, are not only still in existence but form part of the insurgent force NATO is currently combating. This longevity is in no small part due to the transnational support these groups receive due to their sanctuary inside of Pakistan.

In his book "The Hidden War," Artyom Borovik addresses this point bluntly, stating "It is absolutely clear that without the help of the United States, Pakistan, China, and Egypt, the Afghan armed resistance would have had nothing to fight with."[9] In fact, in 1987 alone, 65,000 tons of weapons were purchased by the CIA primarily from China, Egypt, and Israel and then provided to the Mujahedeen via Pakistan.[10] By 1987, at least 80,000 Mujahedeen had undergone training in Pakistan, which frequently consisted of specialized training on explosives, communication, and anti-aircraft systems.[11] To counter the large amount of training and supplies fighters

---

[8] O'Ballance, Edgar. *Afghan Wars: Battles in a Hostile Land, 1839 to Present*. Revised ed. London: Brassey's, 2002. 115-16. Print.

[9] Borovik, ArtemArtyomArtem. *The Hidden War: a Russian Journalist's Account of the Soviet War in Afghanistan*. New York: Atlantic Monthly, 1990. 11. Print.

[10] Yousaf, Mohammad, and Mark Adkin. *Afghanistan--the Bear Trap: the Defeat of a Superpower*. Havertown, PA: Casemate, 1992. 83. Print.

[11] Ibid., 117-119.

flowing from Pakistan received, the Soviets employed brutal scorched earth tactics against insurgents.

This killed many more civilians than insurgents inside Afghanistan, which strengthened the polarization of the Afghan population against the Soviets, but nonetheless exacted a heavy toll on insurgent fighting strength. As part of this, the Soviets launched large offensives against insurgent strongholds within Afghanistan. Specifically, the Soviet attack and insurgent defeat at Zhawar, after a conventional style pitched defense, had the potential to threaten insurgent supply lines providing up to sixty percent of Mujahedeen supplies inside of Afghanistan.[12] The resulting heavy rate of attrition for the Mujahedeen and loss of major supply bases inside Afghanistan worried US military analysts so much they predicted a possible insurgent defeat in the overall war effort.[13]

By 1986, the tide seemed to have turned in the Soviet's favor. The Soviets had cleared major areas, arranged local cease-fires, instituted amnesty plans, and were leveraging the latest advances in satellite surveillance, night vision, and infrared detection systems to mount successful around-the-clock ambushes and attacks against the Mujahedeen.[14] The Soviets significantly refined their tactics and were soon achieving constant harassment of insurgent units by coupling the advanced surveillance assets with the employment of Spetsnaz (Soviet Special Forces) units transported by helicopters.[15] These air assault tactics proved highly effective against insurgents and were continually

---

[12] Ibid., 164.

[13] Ibid., 182.

[14] O'Ballance, Edgar. *Afghan Wars: Battles in a Hostile Land, 1839 to Present*. 154-155.

[15] Galeotti, Mark. *Afghanistan, the Soviet Union's Last War*. London, England: Frank Cass, 1995. 191. Print.

refined, right up until the end of war, when they were assessed at reaching parity with US capabilities and tactics.[16]

The Soviets also demonstrated with some success what this study deems "internal border security." There was a little-noticed event that occurred during the 1982-83 Soviet offensives into the Panjshir Valley that provides evidence of the impact of containing an insurgent using a physical border. The available details are limited, but after laying waste to nearly 80% of the Panjshir Valley, the Soviets erected a six foot high concrete wall at the southern entrance to the valley by the end of 1982.[17]

By the spring of 1983, the already legendary insurgent forces of Ahmed Shah Massoud were forced to broker a yearlong ceasefire to rebuild and resupply.[18] This lasted until the Soviets broke their ceasefire agreement. Although the wall was noted by some to have not been effective, its anomalous occurrence during the war immediately preceding, arguably the single largest ceasefire of the war, is not coincidental when viewed in the context of how border security affects an insurgency.[19] Adding weight to this assertion is the fact that in what amounted to a half-dozen previous Soviet assaults on the Valley, some in division strength, the Soviets did not achieve decisive results or get a ceasefire until they erected the wall.

---

[16] Ibid., 196.

[17] Coll, Steve. *Ghost Wars: the Secret History of the CIA, Afghanistan, and Bin Laden, from the Soviet Invasion to September 10, 2001.* New York: Penguin, 2004. 118. Print.

[18] Urban, Mark. *War in Afghanistan.* Basingstoke: Macmillan, 1988. 118-19. Print.

[19] Coll. 118.

During the early months of 1986, the Soviets launched major offensives against Mujahedeen areas, such as the Mujahedeen base in Zhawar where more than eight thousand insurgents led by Jalalabuddin Haqqani were defeated.[20] (Note again the persistence of some of these groups; the insurgent commander, Haqqani, would reappear leading Taliban elements against NATO forces currently fighting in Afghanistan.) However, due to the ability to fall back into Pakistan where the insurgents could rest and refit, the insurgents proved able to outlast even the most determined of Soviet onslaughts, and immediately reoccupied Zhawar, when the Soviets redeployed their forces to other areas, such as the Kunar and Ghazni Provinces later that year.[21]

This was repeated again and again throughout the war, particularly along the border where the insurgents were the strongest, drawing on the close proximity to arms, supplies, and sanctuary in Pakistan.[22] This was due to the failure of the Soviets to secure the border and cut off the insurgents from their sanctuary. Both the Soviets and the insurgents attest to this explicitly and implicitly in their statements. For example, in an exchange with a Soviet lieutenant colonel, Borovik exclaims that to win, the Soviets would have had to "destroy all of Afghanistan," to which the lieutenant colonel states:

> Nonsense! They should have listened to the military and positioned garrisons along the Pakistani border. If we'd closed all the roads and caravan routes, we would

[20] O'Ballance. 158.
[21] Ibid., 158.
[22] Ibid., 172.

> have squashed the Dukhi (Soviet term for the Mujahedeen) without any military action.[23]

The colonel is not alone in his analysis. Mark Galeotti, in his book, "Afghanistan: The Soviet Union's Last War," argues the war wasn't a military defeat for the Soviets because they never really tried to win. Specifically, he cites "the Kremlin was never prepared to accept the commensurate losses or the political risks of cross-border incursions into Pakistan" to win the war.[24]

Later in his book Galeotti again implicitly concludes the Soviets lost because they were unwilling to deny sanctuary to the insurgents. Specifically, he states that Brezhnev "lacked the political will or theory to swallow his pride and withdraw, let alone pursue the rebels into Pakistan."[25] If any doubt is left to the value of the insurgent sanctuary in Pakistan, Steve Coll's book, "Ghost Wars," could be dedicated in its entirety to this cause. Coll's book is a cover-to-cover testament of the critical importance of Pakistani sanctuary for the Mujahedeen's survival and success.[26]

The Soviet-Afghan War case study provides an excellent opportunity to draw parallels between the US/NATO war efforts and the Soviet efforts, as much of the enemy and terrain remain the same, even if the political reasons for superpower entry into Afghanistan differ greatly. Even a cursory study of Soviet tactical evolution and strategy mirrors the US/NATO strategy in many striking ways. Both forces underwent a rapid evolution of COIN strategy that included reliance on such things as economic revitalization, building

---

[23] Borovik. 236.
[24] Galeotti. 153.
[25] Ibid., 224.
[26] Coll.

projects, training and equipping of the Afghan military, offering insurgents amnesty, and attempts at deals with the insurgents. Many of the same challenges also faced both forces such the constant threat of roadside bombs to vehicles and the chronic ethnic division amongst the population.

Mark Urban, in his book, "War in Afghanistan," titles his last chapter, "Who Will Win."[27] This chapter should be reviewed in its entirety for its almost prophetic parallel to today's contemporary perspective of the trajectory of the US/NATO war in Afghanistan, considering the book was published in 1988 before the actual Soviet defeat in 1989. For brevity, the chapter has been summarized here. Urban begins by noting victory would be defined by the circumstances under which the Soviet army leaves.

Specifically, he links the success or failure of the Soviet backed Afghan government following a military departure, to what will retrospectively be seen as a victory or humiliation of Soviet forces for the entire war effort. He then states the Soviets and their Afghan allies were militarily winning the conflict and had cleared and secured much more area than they originally held in 1980. Bolstering his point, he identifies that the Mujahedeen failed to create a coherent military force, were incapable of taking and controlling even small provincial centers, were unorganized, and could not expand their limited positions. He then points out the success of Soviet-sponsored retraining and equipping of the Afghan forces and how that was a critical factor in increasing Soviet gains. However, Urban does identify that internal support for the Afghan government was limited and growing only slowly. He concludes by

---

[27] Urban. 221-222.

suggesting that it would take another 15 years for the Soviets to conclusively achieve their goals and win, and defeat on the battlefield was highly unlikely.

In the last chapter, Urban identifies why he believed the Soviet "superpower" would be victorious in Afghanistan, but he was proved wrong by the unfolding of subsequent events. To his credit, he did recognize the limited support for the Soviet backed Afghan government and the long term nature of the struggle, but failed to recognize insurgent defeat was not based on what ground they held or even the relative competence of the Afghan military. Considering this, one should quickly ask how this is different, from the US/NATO situation in Afghanistan today. Unfortunately, the answer is the differences are inconsequential and the similarities are clear.

This provides a strong historical pretext to what may lie ahead for the future success or failure of the current US/NATO strategy in Afghanistan. It also suggests most of the current strategy on which US/NATO has pinned its hopes of success, such as training the Afghan military and clearing areas like Kandahar, will prove to be of little consequence. Pakistani Brigadier General Mohammed Yousaf, the former head of Pakistan's Afghan Bureau of the Inter-Service Intelligence (ISI), which was operationally in charge of the Mujahedeen during the war with the Soviets, also notes the Soviet attempts at training the Afghan military were futile in securing the control of Afghanistan.[28]

General Yousaf's remarks referencing the Soviet use of advisors with the Afghan army to provide security against the Mujahedeen should be familiar as the US/NATO force has also invested heavily in a similar

[28] Yousaf. 217.

model in order to stabilize and eventually withdraw from Afghanistan. Further, although Urban did not specifically discuss Soviet action in Kandahar, current operations in that region also have an eerie similarity to the Soviets' experience. Clearing Kandahar was one of the last major Soviet military operations in Afghanistan, and marked almost one year exactly, before the Soviets began their withdrawal in 1988. This correlates very closely with the launch of US/NATO offensives in Kandahar in 2010.

Perhaps, most telling of both conflicts is the continued cross-border, state supported insurgency conducted in an environment absent of effective border security. The similarities between the two conflicts and the historical outcomes should present a strong warning to US military and political leaders, as both the US/NATO forces and the Soviets had predicted military and even political successes, but experienced failures due to the inability to effectively secure the Afghanistan-Pakistan border. Based on this, supporters still claiming the wisdom of the current US/NATO strategy will be dismayed to find out, the most recent US intelligence analysis on the Afghanistan war states explicitly that the war is not going well, and "cannot be won unless Pakistan roots out militants on its side of the border."[29] Although the military rightfully claims to have made gains on the ground in areas such as Marjah, it is incorrect to conclude these gains have any bearing on the overall strategic outcome of the war. The Afghan War Review, is a damning report, and indicates the

---

[29] Associated Press. "Afghanistan, Pakistan Get Bleak Intelligence Brief - FoxNews.com." *FoxNews.com - Breaking News | Latest News | Current News*. 11 Dec. 2010. Web. 16 Dec. 2010. <http://www.foxnews.com/world/2010/12/11/afghanistan-pakistan-bleak-intelligence-brief/>.

US/NATO force is still relying on the flawed assumption Pakistan can be convinced to secure its own border and remove insurgent sanctuary. In summary, if the Soviet superpower was unable to defeat a cross-border rabble, one must concede there should be strong support for the idea that counterinsurgent forces do not typically succeed against state supported cross-border insurgencies.

Chapter Five

Island/Peninsula Paradigm for Border Security in COIN

Anecdotally, from the evidence supporting the benefits of cross-border sanctuary to an insurgent, it appears the natural border containment provided by islands and peninsular countries should offer the counterinsurgent an inherent advantage. Although, there still may be a need to conduct cordoning through internal border security measures, the highly finite contiguous nature of an island state should allow the counterinsurgent to more easily isolate, surround, and destroy and insurgent force.

This seems especially true when a powerful foreign government such as the US is principally conducting and/or supporting the COIN operations on the island or peninsula. Cursory review of the outcome of counterinsurgencies conducted on islands seems to generally support this. The US success in the Philippines, Haiti, and the Dominican Republic provides this evidence.

Further, the British success in Malaysia and the Greek efforts domestically seems to follow this logic. RAND research concurs with the idea island geography was a limiting factor in the Philippines, noting the revolutionary army, as is often the case with island insurgent groups, had a hard time fighting the U.S. Army precisely because of the restrictive archipelagic geography of its homeland. [1] Each island, although large

[1] Rabasa, Angel, Lesley A. Warner, Peter Chalk, Ivan Khilko, and Paraag Shukla. 11.

enough for several military theaters, did not offer much in the way of hiding places, forcing militants to constantly hide among the population after an attack and making it hard to regroup and recuperate without being discovered.[2]

The isolation of the islands, coupled with the U.S. Navy blockade of the archipelago, made it difficult for the insurgents to coordinate actions among themselves.[3] Also, even if the Revolutionary Army had a major outside supporter (which it did not), it would still have been difficult to get support onto the islands, further limiting its chances for victory against the United States.[4]

Following this logic, one would also point to the influence of a strong navy, to provide an effective blockade of insurgent positions, preventing both resupply and retreat as seen in Algeria, Greece, and Israel. Although this theory requires further research, it lends itself well, to the overall findings of this paper. It may have further application to support the viability of leveraging both naval and ground forces in effective COIN strategies to deal with coastal countries such as Somalia and Yemen.

Findings

The hypothesis posed by this thesis withstood analysis through multiple case studies. The hypothesis consistently demonstrated a strong link between successful COIN strategies against cross-border, state supported insurgents and the employment of effective

---

[2] Ibid.
[3] Ibid.
[4] Ibid.

border security. Further, the investigation of situations where COIN strategies against cross-border, state supported insurgencies did not include effective border security, indicated an equally strong link to counterinsurgent failure.

Expanded research also showed a definitive, causal relationship between the use of internal border security strategies, and positive, decisive outcomes against insurgents. Both internal and external border security strategies also proved to have great value beyond COIN operations. Specifically, the Israeli case study demonstrated a clear linkage between effective border security and successful counterterrorism strategies. In fact, in the Israeli case, effective border security proved more valuable in stopping the incidence of terrorist attacks, than the use of intelligence, military operations, and targeted killings combined.

The Israeli study also presents a strong model for the feasibility and effectiveness of unilateral action to secure borders when neighboring populations and states are not just reluctant to support, but overtly hostile. Finally, not only did the hypothesis prove to be valid, but through analysis, a very predictable pattern was identified that applies to both internal and external situations. Specifically, the pattern follows the following general progression:

1. A proto-insurgency begins that, at minimum, has the potential for state sponsorship in a neighboring country.
2. The counterinsurgent begins small scale policing type operations against the "bandits."
3. The counterinsurgent is frustrated by lack of success and the growth of the insurgency.

4. The counterinsurgent begins large scale operations against the insurgency.
5. The insurgent develops a cross-border, state supported sanctuary.
6. The counterinsurgent is repeatedly frustrated in its attempts to decisively engage the insurgency.
7. The counterinsurgent employs border security.
8. The insurgency launches large scale attacks against the border.
9. The insurgency suffers heavy losses in its attempts to breach the border.
10. The insurgency loses its ability to resupply and move its fighters to and from sanctuary.
11. The counterinsurgent begins to achieve decisive gains against the insurgency.
12. The insurgency typically, within a month begins to show signs of critical instability.
13. The insurgency is decisively defeated, neutralized, or seeks a negotiated peace.
14. Political solutions are then achieved over a longer period of time with the state(s) that provided sanctuary to the insurgency, which effectively removes the sanctuary and deals any remaining insurgents a final, decisive defeat.

The inability for the counterinsurgent to achieve decisive effects against an insurgent with a sustained ability to retreat to sanctuary is one of the foundational threats posed by a cross-border, state supported insurgency. The counterinsurgent continuously clears area after area, announces success, and then is dealt a propaganda defeat when the insurgents reappear in strength, sometimes only days later. This undermines the counterinsurgent and reinforces the valid fear amongst the population that the counterinsurgent is unable to protect them.

Even if the population is not supportive of the insurgents, this fear causes the population, to at minimum provide passive support to the insurgency, and at most, actively take part in insurgent operations. Generally, this leads to frustration amongst the counterinsurgent forces and more desperate and violent clearance operations. Aside from the absolutely most disciplined and well trained soldiers, these operations will cause collateral damage, which alienates the population, especially in urban areas where it is extremely difficult to separate the insurgent from the population.

This cycle will repeat itself as long as the insurgency has the ability to withdraw to a sanctuary where it enjoys such necessities as the ability to resupply, draw in recruits, obtain funds, train, and organize. This cycle becomes especially acute and dangerous, when the primary counterinsurgent force is composed of non-native "occupying forces," because it is generally not sustainable for a country to project forces for an extended period of time.

This ultimately leads to the counterinsurgent becoming overextended, exhausted, and defeated. This insurgent strategy has proven so successful against major conventional powers, in places like Vietnam and Afghanistan; it has been dubbed "Fourth Generation Warfare," by authors like Thomas Hammes, in his book, "The Sling and the Stone." However, although these thinkers believe this is the future of conflicts, the addition of border security directly undermines this insurgent strategy and allows the counterinsurgent to again achieve decisive victories.

Analysis has also shown the problem of relying on other states to secure their borders. In short, if the counterinsurgent needs to remove access to a cross-border sanctuary, the counterinsurgent must plan on unilaterally undertaking and accomplishing this task. If the counterinsurgent is unable or unwilling to take on this challenge, the outcome of the struggle will most likely be ultimately decided in the favor of the insurgent. Further, it is only after effective border security is in place that neighboring states harboring insurgents, almost by default due to their own intrinsic interests, will come to the bargaining table.

Although in some cases a country may truly be unable to prevent border incursions by insurgents, in nearly every case of an insurgency that survives beyond the proto phase, the country providing sanctuary does so out of its own interests. Accordingly, reliance on countries like Pakistan is *not* the answer as often claimed in respect to the current US/NATO war in Afghanistan; rather, reliance on these countries is the center of the problem. As such, it is only after the benefits provided to an insurgency have been diminished through effective border security, that a long term and final political decision that decisively ends the insurgency can be achieved. Those that argue the inverse, simply fail to understand the political dynamics of negotiating from a position of strength, vice weakness.

Very ominous is the finding that massive security force assistance (SFA) in the form of weapons (from small arms to heavy weapons and aircraft), training, and provision of advisors, many of whom today are highly trained Special Operations Forces, did not have decisive effects. In fact, although the US/NATO strategy in Afghanistan has placed high priority on SFA and pegged its hopes of victory on standing up a trained local military and police force, there is little historical evidence to suggest this course of action is a game changer. Rather, the evidence shows that even when SFA is massively applied, such as in Vietnam and Afghanistan (by the Soviets), the counterinsurgent still loses. It is only when effective border security is applied, as evidenced in Greece, Algeria, and Israel that the tide is turned against the insurgent. Furthermore, there is substantial evidence to suggest that massive SFA has actually fueled corruption, which undermines the government and provides the insurgency with a steady stream of well-armed and trained recruits and spies.

One of the most difficult hurdles to implementing a full border security strategy is the reeducation of the public and policy makers on the viability and critical necessity of effective border security. The most dangerous consequence of failing to overcome this hurdle is resignation to the thought that even if border security was to be a vital part of counterinsurgency strategy, it "cannot be effectively done" and therefore is neither discussed nor attempted. Proof of this resides in the absence of border security in military doctrine, and perhaps most importantly, in the speeches and writings of leading COIN experts.

One flagrant example is provided by David Kilcullen, a man who has done a great deal toward improving how the US fights insurgencies, but has completely missed the value of border security. After writing chapters on the value of sanctuary and the dangers it poses to the counterinsurgent in "The Accidental Guerilla," he concludes by conceding there is no method to counter it, and pushes the incorrect notion that working with Pakistan is the only solution to fighting the insurgency.

In contrast to the conclusions of these experts, the Greeks, French, Chinese, Israelis, Moroccans, Iraqis, Koreans, and others have all proven securing long borders *is* possible, can be done relatively quickly, and does achieve decisive results.

The rugged mountains of Afghanistan are one of the most frequently cited obstacles to effectively securing the border; however, case study after case study proves mountains are actually one of the easiest types of terrain to secure. Specifically, mountains naturally channel movement to key areas and can be controlled by holding strategic ground.

Yet, skeptics will suggest the mountains in Afghanistan are "more" rugged than other mountainous areas and as such, are not securable. To counter this, one needs to simply reference how the Chinese effectively secured the Himalayan Mountains between China, India, and Nepal to defeat the Tibetan insurgency. There are no mountains on earth higher, more remote, or more rugged than the Himalayas. If these mountains can be secured, the Hindu Kush "foothills" are not the insurmountable problem they are made out to be.

Other critics will still point to cost and manpower, but review of even the most advanced border security systems suggests, the Afghanistan-Pakistan border could be secured for as little as $3.3 billion. This amount is relatively small, compared to the amount of money the US government is spending on the Afghan war effort, especially in regards to the potential return being a decisive victory. Equally important is the actual "return" in reduced US/NATO manpower.

Initially, large numbers of troops would need to be redeployed to border areas, but once secured, the majority of these troops could be replaced with locals, contractors, international forces, or no one at all, once the insurgency ends. The bottom line for policy makers is the recognition that borders, even long and rugged ones, can be effectively secured.

In defense of the current COIN strategy in Afghanistan, border security alone is not sufficient to defeat an insurgency. Although case studies show that a war against a cross-border, state supported insurgent is almost unwinnable without effective border security, the French experience in Algeria proves, winning a COIN struggle requires more than just military might. Considering this, the emergence of rule of law and effective government institutions, efforts to rebuild economies, anti-corruption efforts, and other aspects of what has been termed “soft power” will also continue to be necessary.

COIN operations are not cheap and cannot be won by half measures. Border security is no exception. In fact, the case studies cited in this study, as well as sources not cited (such as the case of Rhodesia), strongly suggest partial employment of border security or conduct of limited, cross-border raids have no decisive effect. Specific to the latter point, any tactical gains achieved by limited, cross-border raids are far outweighed by the strategic costs. Only through effectively securing the border, or the real threat of full scale physical invasion of an area deemed as sanctuary, will decisive results be produced.

Effective border security creates a self-reinforcing situation that overwhelms insurgents. Borders and barriers are a direct affront to insurgent legitimacy and critically hinder their ability to operate. When confronted with effectively secured borders, insurgents must attack and seek to breach these borders, be they internal to a city, or external to physical boundaries of a country. This forces the insurgents to separate from the civilian population where they draw their strength and are able to hide. In turn, this allows the counterinsurgent to decisively bring to bear his superior firepower and destroy the insurgents, without causing collateral damage to the civilian population. The more desperate the situation becomes for the insurgent, the more critical it is for the insurgent to breach the barrier. The more they try, the more the insurgents are interdicted or killed (this paradigm was confirmed as still valid via firsthand accounts from Marines currently operating in Afghanistan). In all case studies, this quickly became an unsustainable situation for the insurgent that directly led to a negotiated settlement or collapse of resistance.

Implementing border security also demonstrates key aspects of COIN such as consistency and commitment to the civilian population. Further, the construction of physical barriers provides needed jobs and processes through which the government can exert its influence and build rapport. The value of this has been highlighted by experts such as David Kilcullen. Once in place, second order benefits to the counterinsurgent will also be realized. Effectively controlled borders foster licit trade networks while limiting illicit networks. This stifling of the black market, specifically in the trafficking of weapons and drugs, directly undercuts the funding for many insurgent groups, warlords, and criminal elements. The existence of these elements directly challenges the legitimacy of the government and must be removed over the long term to bring about a stable functioning state.

Ultimately, effective border security is the *sine qua non* of defeating cross-border, state supported insurgencies. Analysis shows conclusively, although effective border security has been relegated to, at best, a periphery support element in overall COIN strategies, it is in fact a key, critical element that attacks a vital vulnerability and at most, the actual center of gravity of the insurgency. Although successful COIN still requires a holistic approach that simultaneously leverages both hard and soft power to win over the population, provide effective governance, and either win over or eradicate insurgent elements, this is not possible if borders are not effectively secured.

Chapter Six
Policy Implications

The issue of border security in COIN has critical implications for the US/NATO war in Afghanistan and other transnational insurgencies taking root in places like Yemen. It also has a direct impact on the national security of nations around the world that are facing serious cross-border security and terrorist threats. The current US/NATO COIN strategy in Afghanistan, which does not isolate the insurgency from its sanctuary, will be insufficient to achieve the domestic security necessary to neutralize the threat posed to the state by Taliban and Al Qaeda insurgents. Successful COIN in the Afghan War will therefore require a radical change in strategy that includes a concerted effort to seal the Afghanistan-Pakistan border. Otherwise, the United States and NATO risk complete political and military failure in Afghanistan. Further, the US cannot rely on Pakistan as the answer, because it is in Pakistan's interests to continue supporting the Taliban as a hedge against India.

In short, Pakistan *is* the problem. This means the US/NATO must take-on unilateral responsibility for the security of the border. Once established, the conditions that made the Taliban an important national security element for Pakistan to maintain may fundamentally change. Most likely, once unable to operate in Afghanistan, the Taliban will become an internally destabilizing force and threat to Pakistan, who in return will then take decisive action against the Taliban. At this juncture, the US can politically re-engage Pakistan to achieve the long sought political agreements that will decisively end the conflict for the long term.

Based on this study, it is not too late to implement this change. The US - unlike the French in Algeria - has not committed atrocities and as a whole, has not alienated the Afghan population. This means a closure of the border would allow US/NATO forces to consolidate their gains and maintain pacified areas. In fact, this would actually accelerate US/NATO gains and allow decisive actions to occur for the first time. Implementation of a secure Afghanistan-Pakistan border would be devastating to the insurgency because the majority of tier one Taliban fighters (principle agents conducting the war) are foreigners who come from cross-border areas to incite the local populations (tier two Taliban) to arms.

Typically, tier two fighters won't act without the foreign catalyst and certainly would feel more secure working with US/NATO forces. Securing the border would also reign in warlords who challenge the government's legitimacy by financing illicit cross-border smuggling operations. This would include curtailing the drug trade and stifling the movement of bomb making materials, weapons, and ammunition. Further, leveraging locals to help construct a physical barrier would provide critically needed jobs and a process through which the government and legitimate leaders could exercise authority.

The US, more than any other country in the world, excels at large scale construction and engineering tasks. Building a physical border allows the US to decisively put its conventional strengths to work in a COIN environment. For the first time, the US has in hand an effective strategy that not only can win the war, but allow it to regain the initiative against insurgent style warfare worldwide.

The necessity and feasibility of effective border security has policy implications beyond Afghanistan and could be used to effectively address a broad range of issues, such as WMD proliferation, illicit smuggling (weapons, people, drugs), and trans-national terrorism. Border security initiatives also would align better with conditions that have historically been more suitable for United Nations involvement, which could relieve the US of at least some demands in manpower and capital expenditure. UN forces operating on the border between Israel and Lebanon, to maintain the security and integrity of both states' areas, are a contemporary example of this employment.

Border security is also a more politically palatable and effective tool than much riskier strategies, such as covert action and overt military operations, including extensive occupation. For example, providing extensive support to border security operations may be the best option to mitigate terrorist threats emanating from Yemen. This would be unlikely to polarize the population in a way that occupation or discovered covert action does, and could be done jointly with the Yemen government and its neighbors Oman and Saudi Arabia.

Further, border security operations and implementation has a low risk for US personnel, is unlikely to cause any collateral damage, could be used to bolster a nation's economy, and would on the whole be more affordable than many other options. For the United States, this could become the primary policy strategy for dealing with "politically denied areas" where the governments and/or inhabitants of an under-governed or ungoverned area are considered hostile to the United States, such as in Somalia, for the very same reasons it would be effective in a place like Yemen. Thus, helping partner nations implement border security measures to physically contain security threats, may be a more effective, discrete, and less intrusive strategy than other strategies and should become a priority in US policy. In Afghanistan, it may ultimately be the only means of effectively neutralizing the use of Pakistan as a sanctuary for Taliban and Al Qaeda insurgents. Political means have utterly failed to prompt Pakistan to make serious efforts to close its border due to the incompatibility of US goals in Afghanistan with Pakistan's key national security priorities.

The results of this research demonstrate the need to fundamentally alter US military strategy, doctrine, planning, and training. As border security is a key critical element to COIN success, military doctrine related to COIN, specifically Field Manual (FM) 3-24 *Counterinsurgency*, will need to be revised. For example, instead of SFA doctrine focusing on supporting police and military units, it should be revised to target the development of strong border security forces that operate with inherent checks and balances to prevent corruption. General purpose forces (GPF) should also be trained to implement and operate physical border security.

This would include construction of a defense-in-depth, that contains fixed positions on key terrain supported by pre-sited indirect fires and foot, vehicle, and helicopter mobile quick reaction forces that leverage the latest sensors and intelligence, surveillance, and reconnaissance (ISR) assets. Further, at the senior planner level, the specific predicted conditions of a COIN campaign must include an assessment of the nature of any possible insurgency. In the event conditions exist that would suggest a high probability of the development of cross-border sanctuary, the military must plan for extended, comprehensive border security. This will be a paradigm shift in military strategy and planning, and will precipitate the need for training of specific tactics, techniques, and procedures for border security. Until the importance of securing borders when dealing with cross-border insurgencies becomes mainstream within the military, the military will remain blind and unresponsive to this critical element of its overall campaign planning. If there is no guidance for the employment of the force (GEF), no doctrine, no wargaming, and no demand signal from the geographic combatant commanders, the issue will continue to be undervalued or ignored as a necessary COIN activity for success.

In the United States, the ability to implement effective border security means if the US-Mexican border continues to become more violent and dangerous, the US has the means to protect its citizens. The results of this research dispel arguments that the border "cannot" be secured. In fact, the southern US border lends itself well to highly effective border security and interdiction of any infiltrators, as was demonstrated by the Algerian case study where similar terrain was encountered. Domestic border security only remains a matter of political will. Further, the returns in security may far outweigh any costs of implementing a more comprehensive border security initiative.

Conclusion

The results of this research come at a decisive juncture in the US/NATO war in Afghanistan. Should the current course be continued, the US/NATO forces will face a humiliating defeat and a national security disaster. Die hard adherents to the current war plan simply are not getting the information they need to make informed decisions. Acknowledgement of the effectiveness of border security is absent from policy and doctrine and has not, until this book, been systematically studied and addressed in respect to cross-border, state supported insurgencies as faced today in Afghanistan. However, should the US/NATO force incorporate the findings of this research as a key element of a revised strategy, it will have a decisive impact, which will allow the US/NATO force to defeat the insurgency and establish the conditions for Afghanistan to become a sovereign, stable, and functioning state.

These findings fundamentally change the understanding of COIN best practices and have revolutionary implications for US security policy. The results prove border security is the key critical element missing from US COIN strategy and doctrine. Armed with this information, the US can turn its conventional strengths in firepower, industrial might, technology, and manpower to physically securing the Afghanistan-Pakistan border. Only through this action will the US/NATO force turn the tide against asymmetric threats and once again become victorious in insurgency warfare.

Works Cited

Almog, Doron. "The West Bank Fence: A Vital Component in Israel's Strategy of Defense." *The Washington Institute for Near East Policy* (2004): 8. Print.

Anti-Israeli Terrorism, 2006: Data, Analysis and Trends. Rep. Intelligence and Terrorism Information Center at the Israeli Heritage & Commemoration Center, 2007. Web. <http://www.terrorism-info.org.il/malam_multimedia/English/eng_n/pdf/terrorism_2006e.pdf.>.

Associated Press. "Afghanistan, Pakistan Get Bleak Intelligence Brief - FoxNews.com." *FoxNews.com - Breaking News | Latest News | Current News.* 11 Dec. 2010. Web. 16 Dec. 2010. <http://www.foxnews.com/world/2010/12/11/afghanistan-pakistan-bleak-intelligence-brief/>.

Borovik, Artem. *The Hidden War: a Russian Journalist's Account of the Soviet War in Afghanistan.* New York: Atlantic Monthly, 1990. 11. Print.

Cassandra, Adam. "Napolitano: 'You're Never Going to Totally Seal That Border' | CNSnews.com." *CNS News | CNSnews.com.* 25 June 2010. Web. 15 Dec. 2010. <http://www.cnsnews.com/news/article/68494>.

"CBC News - World - Fighting Kills Dozens of Iraqis, 8 U.S. Soldiers." *CBC.ca - Canadian News Sports Entertainment Kids Docs Radio TV.* 6 Apr. 2004. Web. 16 Dec. 2010.

<http://www.cbc.ca/world/story/2004/04/06/iraq_casualties040406.html>.

Chandrasekaran, Rajiv. "Afghan Colonel Vital to U.S. despite Graft Allegations." *Washington Post - Politics, National, World & D.C. Area News and Headlines - Washingtonpost.com.* 4 Oct. 2010. Web. 16 Dec. 2010. <http://www.washingtonpost.com/wp-dyn/content/article/2010/10/03/AR2010100304094.html>.

"CIA - The World Factbook, Algeria." *Central Intelligence Agency.* 9 Nov. 2010. Web. 16 Dec. 2010. <https://www.cia.gov/library/publications/the-world-factbook/geos/ag.html>.

"CIA - The World Factbook, Greece." *Central Intelligence Agency.* 9 Nov. 2010. Web. 16 Dec. 2010. <https://www.cia.gov/library/publications/the-world-factbook/geos/gr.html>.

Coll, Steve. *Ghost Wars: the Secret History of the CIA, Afghanistan, and Bin Laden, from the Soviet Invasion to September 10, 2001.* New York: Penguin, 2004. 118. Print.

Dreazon, Yochi J. "NATO Can't Stop Flow Of Bomb Materials From Pakistan - Friday, October 29, 2010." *NationalJournal.com.* 27 Oct. 2010. Web. 15 Dec. 2010. <http://nationaljournal.com/nationalsecurity/nato-can-t-stop-flow-of-bomb-materials-from-pakistan-20101027>.

Entous, Adam, and Siobhan Gorman. "White House Report Faults Pakistan's Antimilitant Campaign - WSJ.com." *Business News & Financial News - The Wall Street Journal - WSJ.com*. 6 Oct. 2010. Web. 15 Dec. 2010. <http://online.wsj.com/article/SB1000142405274870 3298504575534491793923282.html?mod=WSJ_hpp_ MIDDLENexttoWhatsNewsThird>.

Erased in a Moment: Suicide Bombings Against Israeli Civilians. Rep. USA: Human Rights Watch, 2002. Print.

Erlich, Reuven. "Suicide Bombing Terrorism During the Current Israeli-Palestinian Confrontation September 2000-December 2005." *Intelligence & Terrorism Information Center :: Homepage*. 2006. Web. 16 Dec. 2010. <http://www.terrorism-info.org.il/site/home/default.asp>.

Galeotti, Mark. *Afghanistan, the Soviet Union's Last War*. London, England: Frank Cass, 1995. 191. Print.

Gordon, Michael. "U.S. Begins Erecting Wall in Sadr City." *The New York Times*. 18 Apr. 2008. Web. <http://www.nytimes.com/2008/04/18/world/middle east/18sadrcity.html?_r=4&ref=world&oref=slogin&or ef=slogin>.

Hammes, Thomas X. *The Sling and the Stone: on War in the 21st Century*. Grand Rapids, MI: Zenith, 2006. Print.

Hodge, Nathan. "Gaza Ground Campaign Mirrors Battle of Sadr City? | Danger Room | Wired.com." *Wired.com, Danger Room*. 6 Jan. 2009. Web. 16 Dec. 2010.

<http://www.wired.com/dangerroom/2009/01/gaza-ground-cam/>.

Horne, Alistair. *A Savage War of Peace: Algeria, 1954-1962*. New York: Viking, 1978. 88. Print.
"IDF Spokesperson." Interview. 7 July 2004. Web. <http://www.imra.org.il/story.php3?id=21411>.

Israel High Court Ruling Docket H.C.J. 7957/04: International Legality of the Security Fence and Sections near Alfei Menashe. Supreme Court of Israel. 15 Sept. 2005. Web. <http://www.zionism-israel.com/hdoc/High_Court_Fence.htm>.

"Israeli Security Fence." *Israeli Ministry of Defense*. 2007. Web. 16 Dec. 2010. <http://www.google.com/imgres?imgurl=http://www.securityfence.mod.gov.il/pages/ENG/images/operational1.jpg&imgrefurl=http://www.securityfence.mod.gov.il/pages/ENG/operational.htm&usg=__uSNAFWv_nuvkN9kspZ-gAVCaQ0Q=&h=394&w=680&sz=60&hl=en&start=17&zoom=1&tbnid=P2x_ww6rxtKJ1M:&tbnh=81&tbnw=139&prev=/images?q=israeli+security+fence&um=1&hl=en&sa=N&tbs=isch:1&um=1&itbs=1>.

"Isreal's Security Fence." *Israeli Ministry of Defense*. 2007. Web. <http://www.securityfence.mod.gov.il/Pages/ENG/questions.htm>.

Kershner, Isabel. *Barrier: the Seam of the Israeli-Palestinian Conflict*. New York: Palgrave Macmillan, 2005. 161. Print.

Works Cited

Kershner, Isabel. "Israel Security Fence." *Israeli Ministry of Defense.* 2006. Web. <http://www.securityfence.mod.gov.il/pages/eng/news.htm>.

Kilcullen, David. *The Accidental Guerrilla: Fighting Small Wars in the Midst of a Big One.* Oxford: Oxford UP, 2009. 70-114. Print.

Makovsky, David. "How to Build a Security Fence." *Foreign Affairs* 83 (2004): 50-64. Print.

Nelson, Soraya. "Afghan Border Police Make Progress, Slowly : NPR." *NPR : National Public Radio : News & Analysis, World, US, Music & Arts : NPR.* 5 Mar. 2009. Web. 15 Dec. 2010. <http://www.npr.org/templates/story/story.php?storyId=101377415>.

O'Ballance, Edgar. *Afghan Wars: Battles in a Hostile Land, 1839 to Present.* Revised ed. London: Brassey's, 2002. 115-16. Print.

O'Ballance, Edgar. *The Algerian Insurrection, 1954-1962.* Hamden, CT: Archon, 1967. 39. Print.

O'Ballance, Edgar. *The Algerian Insurrection, 1954-1962.* Hamden, CT: Archon, 1967. 39. Print.

O'Ballance, Edgar. *The Greek Civil War, 1944-1949.* New York: Praeger, 1966. 122. Print.

Prados, John. *Presidents' Secret Wars: CIA and Pentagon Covert Operations since World War II.* New York: W. Morrow, 1986. 164-65. Print.

Rabasa, Angel, Lesley A. Warner, Peter Chalk, Ivan Khilko, and Paraag Shukla. *Money in the Bank: Lessons Learned from past Counterinsurgency (COIN) Operations*. Santa Monica, CA: Rand, 2007. 48. Print.

Radia, Kirit. "New Details on U.S.-Israel Pact to Block Arms to Hamas - Political Radar." *Political Radar*. ABC News, 15 Jan. 2009. Web. 16 Dec. 2010. <http://blogs.abcnews.com/politicalradar/2009/01/new-details-on.html>.

"Siege of Sadr City." *Wikipedia, the Free Encyclopedia*. 14 Nov. 2010. Web. 16 Dec. 2010. <http://en.wikipedia.org/wiki/Siege_of_Sadr_City>.
Thein, Ben. "Is Israel's Security Barrier Unique." *The Middle East Quarterly* Fall (2004). Web. <http://www.meforum.org/652/is-israels-security-barrier-unique#_ftn5#_ftn5>.

Trager, Robert, and Dessislava P. Zagorcheva. "Deterring Terrorism: It Can Be Done." *International Security* Winter 30.3 (2005-2006): 87-123. *Project MUSE*. Web. <http://muse.jhu.edu/login?uri=/journals/international_security/v030/30.3trager.html>.

Tsakalotos, Thrasyvoulos. *Khrónia Stratiótis Tis Ellàdos*. Rep. Vol. Th:40. Athens, 1960. Print. 165-169.
UNSCOB Report, General Assembly. Rep. Vol. A/574. New York, 1948. Print. Para 188.

Urban, Mark. *War in Afghanistan*. Basingstoke: Macmillan, 1988. 118-19. Print.

Works Cited

Van Creveld, Martin. *The Changing Face of War: Combat from the Marne to Iraq*. New York: Ballantine, 2008. 268. Print.

Vitka, William. "Palestinian Militants Attack Border - CBS News." *Breaking News Headlines: Business, Entertainment & World News - CBS News*. 25 June 2006. Web. 16 Dec. 2010. <http://www.cbsnews.com/stories/2006/06/25/world/main1749360.shtml>.
Washington Post 15 Sept. 1963. Print.

Woodhouse, C. M. *The Struggle for Greece, 1941-1949*. London: Hart-Davis, MacGibbon, 1976. 185. Print.
Yousaf, Mohammad, and Mark Adkin. *Afghanistan--the Bear Trap: the Defeat of a Superpower*. Havertown, PA: Casemate, 1992. 83. Print.

Index

Freedom Isn't Free

www.ingramcontent.com/pod-product-compliance
Lightning Source LLC
Chambersburg PA
CBHW020837210326
41598CB00019B/1927
9780997743470